International Futures Programme

The Space Economy at a Glance

2007

ORGANISATION FOR ECONOMIC CO-OPERATION AND DEVELOPMENT

The OECD is a unique forum where the governments of 30 democracies work together to address the economic, social and environmental challenges of globalisation. The OECD is also at the forefront of efforts to understand and to help governments respond to new developments and concerns, such as corporate governance, the information economy and the challenges of an ageing population. The Organisation provides a setting where governments can compare policy experiences, seek answers to common problems, identify good practice and work to co-ordinate domestic and international policies.

The OECD member countries are: Australia, Austria, Belgium, Canada, the Czech Republic, Denmark, Finland, France, Germany, Greece, Hungary, Iceland, Ireland, Italy, Japan, Korea, Luxembourg, Mexico, the Netherlands, New Zealand, Norway, Poland, Portugal, the Slovak Republic, Spain, Sweden, Switzerland, Turkey, the United Kingdom and the United States. The Commission of the European Communities takes part in the work of the OECD.

OECD Publishing disseminates widely the results of the Organisation's statistics gathering and research on economic, social and environmental issues, as well as the conventions, guidelines and standards agreed by its members.

This work is published on the responsibility of the Secretary-General of the OECD. The opinions expressed and arguments employed herein do not necessarily reflect the official views of the Organisation or of the governments of its member countries.

Also available in French under the title:
Panorama de l'économie du secteur spatial

Corrigenda to OECD publications may be found on line at: *www.oecd.org/publishing/corrigenda*.

Foreword

Space technology applications have begun to permeate many aspects of life in our modern societies. They bring substantial improvements to communications, transport, media delivery, weather forecasting, monitoring of the environment and management of natural disasters. With the increasing attention paid to developments in space activities, the need has grown for statistics and analysis in this area to better support and inform policy making.

The Space Economy at a Glance *is an innovative compilation of statistics on the space sector and its contributions to economic activity. As the first-ever OECD statistical overview of the emerging space economy, this book also offers critical insights into some of the main problems involved in deriving internationally comparable data for the industry and its downstream activities. Much remains to be done before being able to develop a universal, data-driven methodology to measure the space sector. This book represents a first step in data collection, as well as providing an initial description of conceptual and definitional differences amongst countries.*

This is the third publication related to space issues produced by the OECD International Futures Programme (IFP). The IFP is a forward-looking multidisciplinary group within the OECD. Its mission is to alert the Secretary-General and the Organisation to emerging issues by pinpointing major developments and analysing long-term concerns in order to help governments map their strategy.

In 2002, in collaboration with the space community, the OECD IFP launched a project to explore how space technologies could potentially contribute to finding solutions to some of the major challenges facing society. Two publications resulted from that in-depth project. Space 2030: The Future of Space Applications *(OECD, 2004) explored promising space applications for the 21st century.* Space 2030: Tackling Society's Challenges *(OECD, 2005) assessed the strengths and weaknesses of the regulatory frameworks that govern space and formulated a policy framework that OECD governments might use in drafting policies to ensure that the potential offered by space is fully realised.*

Upon completion of the two-year project, there was strong encouragement from a number of institutions, especially space agencies, for the OECD IFP to continue exploring the economic dimensions of space infrastructure. In February 2006, the OECD Global Forum on Space Economics was launched. It is an innovative platform for international dialogue on the social and economic aspects of space activities (see Annex A for description). This Forum is supported by contributions from a number of governments and space agencies:

- *ASI (Agenzia Spaziale Italiana, the Italian Space Agency)*
- *BNSC (British National Space Centre)*
- *CNES (Centre National d'Etudes Spatiales, the French Space Agency)*
- *CSA (Canadian Space Agency)*
- *ESA (European Space Agency)*
- *NASA (National Aeronautics and Space Administration)*

- *NOAA (US National Oceanic and Atmospheric Administration)*
- *Norwegian Space Centre (Norsk Romsenter)*
- *USGS (United States Geological Survey).*

The OECD Global Forum on Space Economics aims to provide evidence-based analysis to assist agencies and governments in shaping policy. One of the first tasks the Forum embarked upon was the collection and production of a set of basic data providing a quantitative picture of space-related activities in the OECD area and beyond.

This publication was prepared by Claire Jolly, Policy Analyst, and Gohar Razi, Statistician, in the Secretary-General's Advisory Unit (SGE/AU), under the direction and guidance of Barrie Stevens, SGE/AU Deputy Director and Pierre Alain Schieb, Head of Futures Projects, all working for the OECD Global Forum on Space Economics. Anita Gibson and Belinda Hopkinson of the SGE/AU provided administrative and editorial assistance. This work would not have been possible without the help of Colin Webb, Administrator, Directorate for Science, Technology and Industry (DSTI) and the support of Paul Schreyer and Andreas Lindner of the Statistics Directorate. Thanks also go to Dirk Pilat and Sandrine Kergroach-Connan of DSTI whose earlier paper commissioned by the OECD International Futures Programme was an important building block in the development of this work. Our gratitude as well goes to the organisations participating in the OECD Global Forum on Space Economics.

Table of contents

List of acronyms 11
Executive Summary 13
Introduction 17
Defining the space economy 17
Data sources 18
Structure of *The Space Economy at a Glance* 20
1. Overview of the aerospace sector: background 23
1.1. Size and growth of the aerospace sector – production 24
1.2. Size and growth of the aerospace sector – value added 28
1.3. Aerospace industry research and development 30
2. Readiness: inputs to the space economy 33
2.1. Budgets for space activities 33
2.1.1. Public institutional space budgets 34
2.1.2. Public space research and development budgets 39
2.2 Capital stocks of space assets 42
2.3. Human capital 44
3. Intensity: outputs from the space economy 47
3.1. Revenues from the space industry 48
3.2. Space-related services 50
3.3. International trade in space products 54
3.4. Space patents 56
3.5. Space launch activity 59
3.6. Space exploration-related activities 62
4. Impacts of space activities 65
4.1. Categories of impacts 66
4.2. Commercial revenue multiplier effect for non-space sectors 67
4.3. Impacts on key societal challenges (environment, natural disasters) 69
4.4. Impacts of space programmes on space firms 72
4.5. The way forward 73
5. Spotlights on space activities of selected countries 75
5.1. United states 76
5.2. France 78
5.3. Italy 82
5.4. United Kingdom 86
5.5. Canada 90
5.6. Norway 94

Annex A. **The OECD global forum on space economics** 91
Annex B. **Case study: space technologies and water resources management** 93
The context 93
Role of space systems 94
Investments: The risk-management approach 94
Conclusion 95
Annex C. **General methodological notes** 97
Purchasing power parity (PPP) 97
Production 98
Business expenditure on R&D 98
Current and constant values 98
Nominal and real exchange rates 99
Productivity 99
Double counting 99
Annex D. **Space-related statistics from OECD sources** 101

List of Boxes

1. Classification issues for space activities 19
1.1. The United Nations International Standard Industrial Classification (ISIC) Revision 3.1 Detailed Structure of Class 3530: Manufacture of aircraft and spacecraft 26
2.2. Methodology used to assess the present-day value of the stock of 100 Earth observation satellites (including 20 meteorology satellites) active in 2006 43
3.2. Lessons learned in estimating space-related services revenues: The 2006 UK industry mapping study 52
4.1. Methodological challenges in impacts analysis 66
4.3. Better efficiency due to the use of raw satellite data streams 71
5.2. The space sector in French official statistics 79
Annex B. Tracking the world's water supplies 102

List of Figures

1. Overview of the space economy 18
2. Development of the overall space economy 20
1.1a. Production of aerospace industry in OECD countries, 2003 (or latest year) 25
1.1b. Total aerospace production breakdown in OECD countries, 1980, 1990, 2000 and 2001 25
1.1c. Breakdown of G7 aerospace industry production by year 25
1.1d. Average annual change in aerospace production, 1991-2001 25
1.2a. Value added by aerospace industry for G7 countries, 1980, 1999, 2000, 2001, 2002 29
1.2b. Aerospace value added as percentage of national manufacturing value added for G7 countries, 1980, 1990, 2000, 2001, 2002 29
1.3a. R&D expenditures in aerospace industry by OECD country, 2002 31
1.3b. Aerospace R&D as per cent of manufacturing R&D for selected OECD countries, 1991, 1996, 2002 31
1.3c. BERD of aerospace industry for available OECD countries, 1991, 1996, 2002 31

2.1.1a. Public space budgets as a per cent of national GDP for available OECD and non-OECD countries, 2005 . . . 35
2.1.1b. Countries with operational satellites in orbit as of December 2006 (estimates). 36
2.1.1c. Space budgets of selected OECD and non-OECD Countries, 2005 . . . 37
2.1.1d. US government total space budget, 1990-2007 . . . 37
2.1.1e. Breakdown of total space budgets for OECD countries, 2005 . . . 37
2.1.1f. Breakdown of other OECD space budgets, 2005 . . . 37
2.1.1g. Military as per cent of US total space budget ,1990-2007 . . . 38
2.1.1h. Breakdown of selected European space budgets for space, 2005 . . . 38
2.1.2a. Breakdown of total OECD GBAORD for space, 2004 . . . 40
2.1.2b. GBAORD for space programmes in available OECD and selected non-OECD countries, (latest year) . . . 40
2.1.2c. Space as percentage of national civil GBAORD for OECD countries, 2004 (or latest year) . . . 40
2.1.2d. Space R&D as a percentage of national civilian R&D for selected OECD countries, 1981-2005 . . . 41
2.3a. European space industry productivity and employment, 1992-2006 . . . 45
2.3b. European space industry employment by country, 2006 . . . 45
2.3c. European space industry employment by country and company type, 2005 . . . 45
2.3d. US space industry employment numbers and percentage of total manufacturing, 1997-2004 . . . 45
3.1a. World satellite industry manufacturing revenues, 2000-2006 . . . 49
3.1b. World satellite industry manufacturing revenue by sector, 2000-2006 . . . 49
3.1c. World satellite industry manufacturing revenues by sector, 2000-2006 . . . 49
3.1d. Worldwide launch industry revenues, 2000-2006 . . . 49
3.1e. Worldwide manufacturing of satellite revenues, 2000-2006 . . . 49
3.1f. Turnover by European space manufacturers, 1992-2006 . . . 49
3.2a. World satellite industry revenues for services and other, 2000-2006 . . . 51
3.2b. World satellite services revenue, 2000-2006 . . . 51
3.2c. The three value chains in commercial satellite applications in 2005 . . . 52
3.2d. World government and military commercial satellite market total, 2003- 2012 53
3.2e. World mobile satellite services market: Wholesale and retail revenues, 2003-2012 . . . 53
3.2f. Estimated global expenditures for remote sensing products by application, 2006-2012 . . . 53
3.3a. Amount and share of OECD space products exports, 2004 . . . 55
3.3b. OECD Exports of Space Products 1996-2004 . . . 55
3.4a. Breakdown of space-related patents at EPO, 1980-2003 . . . 57
3.4b. Breakdown of space-related patents granted at USPTO, 1980-2002 . . . 57
3.4c. Breakdown of space-related patenting at EPO, 1980-2004 . . . 57
3.4d. Breakdown of space-related patenting at USPTO 1980-2004 . . . 57
3.4e. Breakdown of space-related patents by type and country at EPO, 1980-2004 . . . 58
3.4f. Breakdown of Space-related Patents by Type and Country at USPTO, 1980-2004 . . . 58
3.5a. Total commercial and non-commercial launch events 1998-2006 . . . 60
3.5b. Total worldwide commercial and non-commercial payloads, 1998-2006 . . . 60
3.5c. Total (commercial and non-commercial) launch events by country, 2000-2006 60

3.5d. Breakdown of 177 worldwide commercial launch events, 1996-2000 61
3.5e. Breakdown of 111 worldwide commercial launch events, 2001-2006 61
3.5f. Total worldwide commercial launch events and revenue, 1997-2006 61
4.2. Impacts of US commercial space transportation and enabled industries, 2004 68
4.3a. Number of people affected per disaster type 70
4.3b. Economic and insured losses due to disasters: Absolute values and long-term trends, 1950-2005 70
4.4. Norway space industry spin-off factor, 1997-2005, with company forecasts up to 2009 72
5.1a. US space manufacturing industry 77
5.1b. Contribution of space industry employment and value addedto US economy, 1997-2004 77
5.1c. Space *versus* manufacturing value added growth rate, 1998-2004 77
5.1d. US satellite telecom revenue and percentage of telecom revenue, 1998-2004 77
5.2a. Evolution of the French space manufacturing turnover by type of activities unconsolidated 80
5.2b. Space and aeronautics as Percentage of Turnover of 221 active firms in the aerospace sector in the Midi Pyrenees region, 2004 (%) 80
5.2c. Space and aeronautics shares in aerospace firms turnover in Midi-Pyrénées and Aquitaine, 2004 81
5.2d. Revenues for satellites and related space systems by applications 81
5.3a. Breakdown of Italian space enterprises by sector 83
5.3b. Breakdown of Italian space enterprises by activity/skill type 83
5.3c. ESA contracts to Italy per directorate 84
5.3d. Employment in Italian Space Industry by industry type, 2005 84
5.3e. Employment in Italian Space Industry by industry type, 2005 85
5.4a. UK space industry upstream and downstream real turnover, 1999-2005 87
5.4b. Breakdown of UK upstream turnover, 2004/05 87
5.4c. Breakdown of UK downstream turnover, 2004/05 88
5.4d. Turnover of UK space industry customers by region and type, 2004/05 88
5.4e. Breakdown of UK turnover by application, 2004/05 89
5.5a. Canadian space sector revenues and employment, 1996-2005 91
5.5b. Canadian space sector revenue breakdown, 1996-2005 92
5.5c. Canadian space sector domestic revenue breakdown, 1996-2005 92
5.5d. Canadian space sector export revenue source breakdown, 1996-2005 92
5.5e. Canadian space sector total revenue by categories, 1996-2005 93
5.5f. Canadian space sector revenue by activity sector, 1996-2005 93
5.6a. Turnover of Norwegian-produced space goods and services, 1997-2009 95
5.6b. Export share as percentage of total Norwegian space-related turnover, 2002-2005 95
5.6c. Spin-off effects factor for Norwegian ESA and NSC contracts, 1997-2009 95
5.6d. Total Norwegian ESA contracts and non-ESA spin-off sales, 2005 95

List of Tables

1.1. State of the aerospace sector in 2005-2006 in selected countries 27
2.2. Estimated annual world infrastructure expenditure (additions and renewal) or selected sectors, 2005, in USD 43
3.3. G7 total exports of space products, 2004 55
3.6a. Selected active and upcoming robotic exploratory probes, as of December 2006 63
3.6b. Selected human spaceflight statistics as of December 2006 63
4.1. Selected types of impact of space investments........................... 66
4.2a. Economic impacts of the US commercial space transportation and enabled industries, 2004 (thousands of USD) 68
4.2b. Economic impacts (revenues and jobs) throughout major US industry sectors, generated by commercial space transportation and enabled industries, 2004 . 68
5.2a. Turnover from manufacturing of launchers and space vehicles in France in 2005 (NAF code: 35.3C) 79
5.2b. Evolution of French space manufacturing turnover, per activity and total 80
Annex B. Main evaluation methods for the analysis of large programmes 96

List of Acronyms

AIA	Aerospace Industry Association of America
AIAC	Aerospace Industries Association of Canada
AIAD	Associazione delle Industrie per l'Aerospazio i Sistemi e la Difesa (Italian Industry Association for Aerospace Systems and Defence)
AIPAS	Associazione Italiana PMI per l'Aerospazio (the Association of Italian Small and Medium Aerospace Enterprises)
ANBERD	OECD Analytical Business Enterprise Research and Development database
ASAS	Associazione per i Servizi, le Applicazioni e le Tecnologie ICT per lo Spazio (Association for Space-based Applications and Services)
ASD	European AeroSpace and Defence Industries Association
ASI	Agenzia Spaziale Italiana (Italian Space Agency)
BTD	The OECD STAN Bilateral Trade database
BERD	Business Enterprise Research and Development
BNSC	British National Space Centre
CAD	Canadian dollars (currency code)
CAST	China Academy of Space Technology
CNES	Centre National d'Études Spatiales (French Space Agency)
COMTRADE	United Nations' Commodity Trade Statistics Database
CPI	Consumer Price Index
CSA	Canadian Space Agency
CSG	Centre Spatial Guyanais (Space Centre in French Guiana)
DARS	Digital Audio Radio Services
DBS	Direct Broadcast Satellite
DTH	Direct to Home satellite
EO	Earth Observation satellite
EPO	European Patent Office
ESA	European Space Agency
ESA95	European System of Accounts 1995
ESTEC	European Space Research and Technology Centre
ESTP	European Space Technology Platform
EU	European Union
EUR	Euro (currency)
EUROSTAT	Statistical Office of the European Communities
FAA	US Federal Aviation Administration
FAI	Fédération Aéronautique Internationale (International Aeronautic Federation)
FAA/AST	US Federal Aviation Administration's Office of Commercial Space Transportation
FSS	Fixed Satellite Services

G7	Group of 7 leading industrial nations (Canada, France, Germany, Italy, Japan, United Kingdom, United States)
GBAORD	Government Budget Appropriations or Outlays for R&D
GBP	British pounds (currency code)
GDP	Gross Domestic Product
GIFAS	Groupement des Industries Françaises Aéronautiques et Spatiales (French Aerospace Industries Association)
GPS	Global Positioning System
IAF	International Aeronautic Federation
ICT	Information and Communication Technology
IFP	OECD International Futures Programmes
INSEE	Institut National de la Statistique et des Études Économiques (French National Institute for Statistics and Economic Studies)
IPC	International Patent Classification
ISIC	United Nations International Standard Industrial Classification
ISS	International Space Station
ITCS	International Trade in Commodity Statistics Database (UN/OECD)
JAXA	Japan Aerospace Exploration Agency
MSS	Mobile Satellite Services
NACE	Nomenclature d'Activité dans la Communauté Européenne (Economic Classification System in the European Community)
NAF	Nomenclature d'Activités Française (Economic Classification System in France)
NAICS	North American Industry Classification System
NASA	US National Aeronautics and Space Administration
NOAA	US National Oceanic and Atmospheric Administration
NOK	Norwegian krone (currency code)
NSC	Norwegian Space Centre (Norsk Romsenter)
OECD	Organisation for Economic Co-operation and Development
PPP	Purchasing Power Parities
R&D	Research and Development
RIMS II	Regional Input-Output Modelling System II
ROI	Return on Investment
SBAC	Society of British Aerospace Companies
SESSI	Service des études et des statistiques industrielles (French Service for Industrial Studies and Statistics)
SIA	US Satellite Industry Association
SNA	UN System of National Accounts
SOHO	Solar and Heliospheric Observatory (ESA and NASA)
SSB	Space Studies Board (US)
STAN	OECD Structural Analysis Statistics database
UK	United Kingdom
UN	United Nations
US	United States
USD	US dollars (currency)
USGS	US Geological Survey
USPTO	United States Patent and Trademark Office
VSAT	Very Small Aperture Terminals

THE SPACE ECONOMY AT A GLANCE – ISBN 978-92-64-03109-8 – © OECD 2007

Executive Summary

Space applications have the potential to provide significant contributions to society's responses to 21st century challenges, such as environmental monitoring, management of natural resources, security and safety. Key activities in everyday life – weather forecasting, global communications and broadcasting, disaster prevention and relief – depend increasingly on the unobtrusive utilisation of space technologies. Over coming decades, space-related applications, such as distance education, telemedicine, precision farming, land use management, and monitoring of various international treaties, will continue to hold great socio-economic promise.

In order to ensure that the potential of space is more fully realised, governments and space agencies need evidence-based analysis to help shape policy making. Paradoxically, despite the critical role that the space industry plays in modern society, the space sector is one of the least developed in terms of robust, internationally comparable statistics and data.

The Space Economy at a Glance responds to this growing need for metrics by assembling a basic data set that gives a quantitative picture of space-related activities in OECD countries and several major non-OECD economies. The book also offers critical insights into some of the main problems involved in deriving internationally comparable data for the industry and its downstream activities, notably the lack of detailed data, and conceptual and definitional problems.

What is the "space economy"?

In 2006, the OECD International Futures Programme launched the OECD Global Forum on Space Economics – a platform for international dialogue and research amongst participating national governments and space agencies – and set out to explore further the economic dimensions and societal impacts of space infrastructure and space-based activities. The Forum defines the emerging space economy as:

> All public and private actors involved in developing and providing space-enabled products and services. It comprises a long value-added chain, starting with research and development actors and manufacturers of space hardware (*e.g.* launch vehicles, satellites, ground stations) and ending with the providers of space-enabled products (*e.g.* navigation equipment, satellite phones) and services (*e.g.* satellite-based meteorological services or direct-to-home video services) to final users.

How is the "space economy" measured?

An increasing number of countries are developing space systems and applications, but internationally agreed definitions for statistical terminology on space activities do not yet exist. The current edition of the United Nations International Standard Industrial Classification (ISIC Revision 3.1) includes most parts of the space sector under different aggregate categories. Indeed, there is no specific "space activity" classification in the ISIC, and disentangling the space sector from the larger aerospace sector remains a challenge in most countries.

Inspired by many years of OECD work on emerging economic areas (*e.g.* the information, society, e-commerce, the bio-economy), the data and statistics on the space economy are presented here in a framework that consists of: readiness (inputs, such as financial and human resources), intensity (outputs, such as products and services), and impacts (largely qualitative, societal "value added").

This book compiles information on space-related manufacturing, goods and services, public budgets, R&D, human capital, and patents from a wide range of official and non-official sources. Official statistics come from two main sources: OECD databases and publications (such as the OECD Structural Analysis or STAN system of databases), and official government departments or national space agencies. Non-official sources are mainly industry associations and private consulting firms.

The compiled data answer questions, such as:

- Who are the main space-faring nations?
- How large are revenues and employment in the sector?
- How much R&D goes on, and where?
- What is the value of spin-offs from space spending?

Spotlights on space developments in some members of the OECD Global Forum on Space Economics provide country-specific information for the United States, France, Italy, the United Kingdom, Canada and Norway. And highlights from a recent internal OECD study of water resources management illustrates the vital role that space applications can have in providing innovative solutions.

Some statistical findings

Estimates of the size of the space economy vary widely (due to the lack of internationally comparable data). However, worldwide, institutional budgets (around USD 45 billion in 2005 for OECD countries) and new commercial revenues from space-derived products and services (around USD 110-120 billion in 2006 worldwide) indicate that the underlying trend in the space economy is one of growth.

The G7 countries dominate production in the aerospace industry, which comprises the manufacture of all aircrafts and spacecrafts. Although a key strategic sector for many countries, aerospace represents a small share of their total manufacturing value added, ranging in 2002 from below 1% in Japan to more than 3% in Canada, France, the United Kingdom (UK) and the United States (US). Business enterprise R&D expenditure in the aerospace sector totalled more than USD 19.8 billion in 2002; the US, France, the UK and Germany accounting for 84% of that total.

THE SPACE ECONOMY AT A GLANCE – ISBN 978-92-64-03109-8 – © OECD 2007

Narrowing to the smaller space sector, downstream space activities (applications) are often much larger than the traditional upstream segment (manufacturing). In 2006, manufacturing revenues (*e.g.* satellites, rockets) were estimated at around USD 12 billion and space-related services (*e.g.* direct to home satellite television, GPS) were estimated at more than USD 100 billion. As for human resources in the space industry, data are very fragmented; but an estimated 120 000 people in OECD countries are employed in upstream sectors in 2006.

Capital stocks, as well as annual levels of investment, for space assets are very difficult to estimate; however focusing on satellites' values, a 2005 study estimated that the 937 satellites in the Earth's orbit at that time had a replacement value of USD 170 to 230 billion.

Finally, patent data are considered an indicator of technological innovation and the economic vigour of a given sector. Between 1990 and 2000, the number of space-related patents tripled both in Europe and the US, with the US, France, Germany and Japan accounting for a major portion.

Broad socio-economic impacts

Countries develop space activities for both political (*i.e.* international prestige) and strategic objectives (*i.e.* civilian-military utilisation of space systems). Key outputs of those activities comprise various scientific and technological developments (*e.g.* space exploration, advances in physics), even as unforeseen socio-economic impacts can increasingly be detected in the larger economy.

The many derived space-based services have positive impacts on economies and societies, although at this stage, they are more qualitative than quantitative. The ability to disseminate information over broad areas, instantaneous telecommunications, and a global vision of the world are some of the important capabilities that space assets bring. Combining terrestrial facilities with space infrastructure can provide benefits for end users such as: decreased transaction time, cost savings, cost avoidance, improved productivity, and increased efficiency.

Studies show that being able to transfer and broadcast information worldwide instantaneously has been a significant commercial revenues' multiplier since the 1980s for phone and television companies. Employment in the space sector has led to the creation of jobs in "derived" sectors, in particular telecommunications.

In Norway, the "spin-off effect" of space programmes on space firms has been measured at 4.4, that is for every million Norwegian kroner (NOK) of governmental support, space sector companies have on average attained an additional turnover of NOK 4.4 million (EUR 510 000). Although this impact measure may vary widely depending on the country and level of specialisation, it is indicative of possible increased competitiveness due to space involvement.

Benefits from space infrastructure are becoming more evident in the management of long-term and significant challenges of modern society. In the case of natural disaster management (floods, for example), remote sensing from space can provide data for the whole cycle of information for flood prevention, mitigation, pre-flood assessment, response (during the flood), recovery (post-flood) and weather newscasts. Timely satellite imagery and communications links in hard-to-reach places can help stem catastrophic loss of lives and economic losses.

Challenges to overcome in data collection

Future space-related data collection efforts will need to overcome obstacles in order to more accurately quantify the space sector and render data and statistics comparable across countries. The challenges include:

- Disaggregating data. Disentangling space data from aircraft data in the larger aerospace sector will be essential; likewise, separating manufacturing data from services, in some instances.
- Double-counting. Production data in the sector is often subject to double-counting. Efforts to break out value added will be necessary.
- Limited international comparability. Countries are using their own methodologies, concepts, and definitions in official government data.
- Confidentiality. Much data is subject to secrecy due to dual-use military and civilian applications of space developments and/or the existence of only one or a few major space-related companies in a country.
- Non-OECD countries. As in the case of other economic sectors, obtaining official data is difficult, and purchasing power parity issues need to be taken into account.
- Employment. Data is not available split by R&D or production, for instance.
- Detailed services. Only satellite telecommunications services have been partly traced; trade in other services is poorly quantified.

Next steps

Much work remains to be done to develop universal, data-driven indicators for the emerging space economy. More efforts in that direction could benefit both decision-makers, industry and citizens, and help them have a better understanding of the significance of space activities in the larger economy.

Further actions could include international efforts to separate the statistical classifications for aircraft and spacecraft industries, as well as exercises that drill down on space-related services (such as telecoms, satellite navigation). Case studies that assess the social and economic impacts of space applications in today's world would help to better qualify and quantify the space economy. The OECD Global Forum on Space Economics could be the platform that provides the impetus for such work, while further international co-operation will be required with national statistical offices, space agencies and industry associations.

Introduction

The Space Economy at a Glance breaks new ground in a number of ways. Prepared under the aegis of the OECD Global Forum on Space Economics, this publication constitutes a first attempt at providing a quantitative, internationally comparable view of not only the space sector itself, but also its broader role in the economy and society. It brings together data and statistics from official and unofficial sources that cover public space budgets, space sector revenues, trade in space products, and space patents, in order to illustrate the economic and societal impacts of space-based activities. While some of the statistics are drawn from traditional sources in the space community, others are drawn from OECD databases, containing in some cases previously unpublished material.

It is important to note that, as with all new emerging economic sectors, official statistics in the domain of commercial space activities are considerably underdeveloped. This is because more detailed data often require new or revised statistical definitions and classifications. In order to improve international comparability of such data, significant international co-operation will be needed. Looking to the future, the OECD Global Forum on Space Economics could provide a platform for such work.

1. Defining the space economy

Space technologies are becoming an increasingly important part of everyday life. Weather forecasting, air traffic control, global communications and broadcasting – these and many other essential activities would be almost unthinkable today without satellite technology. Although an increasing number of countries are developing space systems and applications, internationally agreed definitions for statistical terminology on space activities do not yet exist.

In broad terms, the OECD Global Forum on Space Economics defines the space economy as:

> All public and private actors involved in developing and providing space-enabled products and services. It comprises a long value-added chain, starting with research and development actors and manufacturers of space hardware (*e.g.* launch vehicles, satellites, ground stations) and ending with the providers of space-enabled products (*e.g.* navigation equipment, satellite phones) and services (*e.g.* satellite-based meteorological services or direct-to-home video services) to final users.

Thus, the space economy is larger than the traditional space sector (*e.g.* rockets and launchers); and it involves more and more new services and product providers (*e.g.* geographic information systems developers, navigation equipment sellers) who are using space systems' capacities to create new products. Figure 1 provides a simplified view of the space economy; a public or private actor may be involved simultaneously in several space activities (*e.g.* being a manufacturer, as well as an operator and service provider).

As a first step in quantifying the space economy, this publication focuses mainly on the data available for the traditional space sector; however, throughout the document, broad indications about derived sectors of the space economy will be provided. More methodological work is needed to capture in greater detail space economy-related services.

Figure 1. **Overview of the space economy**

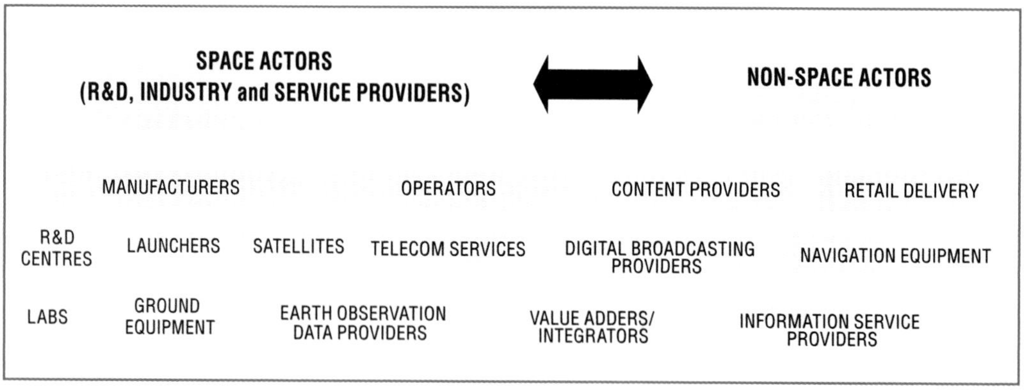

SPACE ACTORS
(R&D, INDUSTRY and SERVICE PROVIDERS)
NON-SPACE ACTORS
MANUFACTURERS
OPERATORS
CONTENT PROVIDERS
RETAIL DELIVERY
R&D CENTRES
LAUNCHERS
SATELLITES
TELECOM SERVICES
DIGITAL BROADCASTING PROVIDERS
NAVIGATION EQUIPMENT
LABS
GROUND EQUIPMENT
EARTH OBSERVATION DATA PROVIDERS
VALUE ADDERS/ INTEGRATORS
INFORMATION SERVICE PROVIDERS

Source: OECD IFP (2006).

Governments play a key role in the space economy as investors, owners, operators, regulators and customers for much of space infrastructure. As in the case of other large infrastructure systems (*e.g.* water, energy), government involvement is indispensable to sustain the overall space economy and to deal with strategic implications of such complex systems. In the case of space, infrastructure can be used for both civilian and military applications as space technologies are by nature dual use, and military developments often pave the way for the development of civil and commercial applications (*i.e.* today's rockets are derived from missiles).

Estimates of the size of the space economy vary considerably, due the lack of internationally comparable data. Worldwide, institutional budgets (around USD 47 billion in 2005 for OECD countries) and new commercial revenues from space-derived products and services (around USD 110-120 billion in 2006) indicate that the underlying trend in the space economy is one of growth. And this remains true, despite the cyclical nature of commercial space activities (*e.g.* regular replacement of telecommunication satellite fleets).

2. Data sources

While focusing on OECD countries, *The Space Economy at a Glance* also looks at selected international non-OECD countries in the space economy, using both official government and private (association or industry) statistics. A major methodological challenge concerns the statistical classifications for space activities, which suffer from a lack of clear and common official definitions (see Box 1).

Box 1. **Classification issues for space activities**

The ISIC – The United Nations International Standard Industrial Classification (ISIC) system is a standard classification of economic activities arranged so that entities can be classified according to the activity they carry out. The current edition of the ISIC (Revision 3.1) includes most parts of the space sector under different aggregate categories. This will remain case in the forthcoming (2007) international ISIC (Revision 4.0).

"Space" in ISIC – There is no specific "space activity" classification in the ISIC. Most national industrial classifications (based largely on the UN ISIC system) used by statistical offices provide no breakdown for this industry. However, the US North American Industry Classification System (NAICS) and the French national statistical system (NAF) are partial exceptions. Nationally and regionally, countries have adopted the main international categories, while sometimes adding more details in how they classify industries. Three ISIC aggregate divisions cover the main space sector activities:

Division 35: Manufacture of transport equipment / Class 3530: Manufacture of aircraft and spacecraft. This class includes the manufacture of spacecraft and spacecraft launch vehicles, satellites, planetary probes, orbital stations and shuttles.

Division 62: Air transport / Class 6220: Non-scheduled air transport. This class includes the launching of satellites and space vehicles, and the space transport of physical goods and passengers.

Division 64: Post and telecommunications / Class 6420: Telecommunications. This class includes the transmission of sound, images, data or other information via satellite.

Other segments of the space sector, especially space applications and services, are even more "buried" within official statistical classifications. For example, ground equipment and communication equipment are included under broader categories in the manufacturing sector.

"Space" in the next ISIC (2007) – Overall, the space sector will not have much more visibility in the 2007 revision of the ISIC (Revision 4.0), although satellite communications activities will be better represented:

Manufacturing in the aerospace sector will still be counted as a single activity in Class 303: Manufacture of air and spacecraft and related machinery (within Section C: Manufacturing, and Division 30: Manufacture of other transport equipment).

The satellite communications domains will, however, be more precisely represented. Class 6130: Satellite telecommunications activities (within Section J: Information and communication, and Division 61: Telecommunications) will include the activities of "operating, maintaining or providing access to facilities for the transmission of voice, data, text, sound, and video using a satellite telecommunications infrastructure." This class will include the delivery of image, sound or text programming received from cable networks, local television stations, or radio networks to consumers via direct-to-home satellite systems. This class will also take into account the provision of Internet access by operators of satellite infrastructures.

Source: UN Statistics Division, http://unstats.un.org/, accessed 6 May 2006.

Official statistics in this book consist of data from two main sources, OECD databases and publications, and official government departments or national space agencies:

- The primary OECD source is the STAN (or "Structural Analysis") system of databases that were essential for compiling comparable official statistics for industrial and national variables. Other OECD data sources include the UN/OECD International Trade in Commodity Statistics database (ITCS), the Annual National Accounts, and selected Exchange Rates and Purchase Power Parities databases.
- Official data also include reports and documents from government departments and agencies (*e.g.* national space agencies).

Private sources of data in this book include industry associations and consulting firms. While the data from those sources are quite comprehensive, all raise questions with respect to their international comparability.

3. Structure of *The Space Economy at a Glance*

Inspired by many years of OECD work on emerging economic areas (*i.e.* the information society, e-commerce, the bio-economy), the data on the space economy are presented in a framework that consists of three stages: readiness (inputs), intensity (outputs), and impacts. Each stage provides an indication of the maturity of the sector. The diagram below (Figure 2) illustrates the different steps from readiness to impacts. This is, of course, a stylised representation, as some space applications (*e.g.* satellite telecommunications) are more developed than others and are already making a very significant impact.

Figure 2. **Development of the overall space economy**

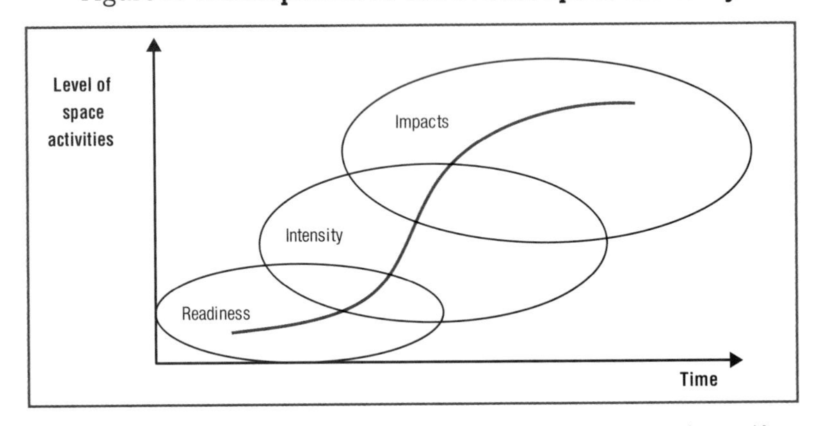

Source: Figure adapted from the OECD *Working Party on Indicators for the Information Society, Guide to Measuring the Information Society*, Directorate for Science Technology and Industry, Committee for information, computer and communications policy, 8 November 2006, DSTI/IICP/IIS(2005)6/FINAL, p. 10.

This publication has five chapters: (1) an overview of the aerospace sector, historically the cradle for space activities; (2) the readiness (inputs) of the space economy, (3) its intensity (outputs), (4) its impacts, and (5) several spotlights on the space programmes of selected countries. The annexes provide methodological notes, tables with the underlying data for some OECD graphs, and information on the OECD Global Forum on Space Economics and a list of participating governments and space agencies.

1. *The background on the aerospace sector* provides the wider context from which the space economy has emerged. It also highlights the importance of future endeavours to separate the aircraft and spacecraft industry components for more meaningful official OECD data.
2. *The readiness factors (inputs) of the space economy* consist of the overall technical, commercial, and financial infrastructures necessary to engage in pertinent space activities. This chapter deals with the financial and human resources that are employed in producing space-related hardware and the provision of relevant services. It examines R&D, financial support for space programmes, and human capital.
3. *The intensity factors (outputs) of the space economy* describe the use that is made of space activities. The outputs refer to the specific space-related outcomes that are derived from the inputs. Thus, outputs may include products or services that are produced or provided in the realm of the space sector. They also include the benefits to industries/ nations, including financial benefits (sales and trade revenues) and indications of future financial benefits (*i.e.* patents).
4. *The impacts of the space economy*, which are more qualitative than quantitative, consist of the "societal value-added" created by space activities. Examples provided are of benefits to society as a whole.
5. *The spotlights on selected countries* offer some insights into the space-related activities of member countries participating in the OECD Global Forum on Space Economics. Data come from their official sources (such as national space agencies or statistical offices) as well as private sources. Direct comparisons between countries are not possible due to definitional, conceptual and methodical differences.

The quality of available measures and comparable data for the space economy varies strongly for the input, output and impact stages. Some official statistical data are available for the readiness (input) factors (although not always readily comparable) and the intensity (output) factors, but these need to be supplemented by private data sources (*e.g.* industry surveys for revenues of the space sector). There are very few data on impacts. This situation is mainly due to a current lack of comparable quantitative information internationally. In order to provide a better indication of the state of the space economy, more work on the concepts and definitions for the space sector and the larger space economy will be needed. This will call for significant international co-operation. Looking to the future, the OECD Global Forum on Space Economics could provide a platform for such work.

1. OVERVIEW OF THE AEROSPACE SECTOR: BACKGROUND

The space economy evolved from the aerospace industry and the two still share many aspects, components and technologies (e.g. space launchers are modified guided missiles). Detailed examination of the space sector is hampered by this legacy since many data are still classified according to categories defined for aerospace. As the UN International Standard Industrial Classification (ISIC) summarised in Box 1.1 shows, this covers everything from hang gliders to space shuttles, so national statistical offices and space agencies are working to make a clearer distinction between space and aerospace classifications. This will enhance the availability and accuracy of data on the space economy in the future.

The present classification is useful though for establishing the space economy in the wider context of the aerospace sector, and the following sections examine trends in aerospace production, value added, and business research and development. The OECD data are based on the broad aggregated ISIC Class 3530: Manufacture of aircraft and spacecraft.

1.1. SIZE AND GROWTH OF THE AEROSPACE SECTOR – PRODUCTION

According to ISIC 3530, the aerospace industry comprises the production of all aircraft and spacecraft, but space-related services such as telecommunications are not included. They may however be indirectly reflected in aerospace production as intermediate inputs. This section explores the aerospace sector by examining the size and growth of the industry in OECD countries.

Highlights

The largest aerospace producer in 2003 was the US (USD 126 billion), followed by the rest of the G7 (see Figure 1.1a). Italy (the smallest G7 aerospace producer) manufactured twice as much as Spain, the leading non-G7 producer.

OECD production since the 1980s shows two trends (see Figure 1.1b):

1. Value has been rising.
2. Domination of OECD aerospace by the G7 countries is not a new phenomenon, but relative performances have changed. In 2001, for example, 95% of OECD production came from the G7, but over the past two decades, G7 production has shifted away from the US and Italy towards France, Germany and Canada (see Figure 1.1c). Data from 1991-2001 show that production in France and Canada grew by over 10% on average, while that of the US and Italy varied little (Figure 1.1d).

After a commercial downturn in the late 1990s, data for 2005 and 2006 reported by national industry associations show significant growth in revenues (since 2004), supported largely by global demand for air transport and defence equipment (Table 1.1)

Definition

The aerospace industry refers to Class 3530 of the United Nations' International Standard Industrial Classification (ISIC) Revision 3.1 which covers the manufacture of aircraft and spacecraft. This broad class comprises the manufacturing of both non-space items (passenger and military aeroplanes, helicopters, gliders, balloons, etc.) and space items (including spacecraft, spacecraft launch vehicles, satellites, planetary probes, orbital stations and shuttles). This also includes the manufacturing of their parts and accessories, used in civil or military applications. Production refers to the total value of this class of goods produced in a year, whether sold or stocked.

Methodology

Production includes intermediate inputs (such as energy, materials and services required to produce final output). An item produced in this industry may show up twice in production: as the final output of one enterprise and the intermediate input of another one. This means that double counting could potentially be a problem.

Data comparability

The data here come from OECD's Structural Analysis Statistics (STAN) database, which includes statistics for all OECD countries (except Turkey). Countries that either had no or zero values were Austria, the Czech Republic, Denmark, Greece, Ireland, Luxembourg, New Zealand, Portugal, the Slovak Republic and Switzerland. German data prior to 1991 only cover Western Germany. Although some statistics are actual country submissions, most are estimates based on surveys or other data from the OECD member countries. To put the values into a common measure, Purchasing Power Parities (see Annex for details on PPP) were used to convert current production values into US dollars.

Data sources

- AeroSpace and Defence Industries Association (ASD) (2007), ASD Industry Figures 2006: The Status of the European Aerospace and Defence Industry 2006, June.
- Aerospace Industries Association of Canada (AIAC) (2006), *Performance Results for Canada's Aerospace Industry in 2005*, July.
- Associazione delle Industrie per l'Aerospazio i Sistemi e la Difesa (AIAD) (2007), website *www.aiad.it*, February.
- Groupement des Industries Françaises Aéronautiques et Spatiales (GIFAS) (2007), *French Aerospace Review 2006*, 16 March.
- OECD (2007), OECD Structural Analysis Statistics, STAN Industry database, OECD, Paris, April.
- OECD and Eurostat (2007), Purchasing Power Parities and Real Expenditures: 2002 Benchmark Year 2004 Edition, OECD, Paris, April.
- Society of British Aerospace Companies (SBAC) (2007), *UK Annual Aerospace Survey*, June.

1.1. SIZE AND GROWTH OF THE AEROSPACE SECTOR – PRODUCTION

Figure 1.1a. Production of aerospace industry in OECD countries, 2003 (or latest year)

Billions of current US dollars using PPP

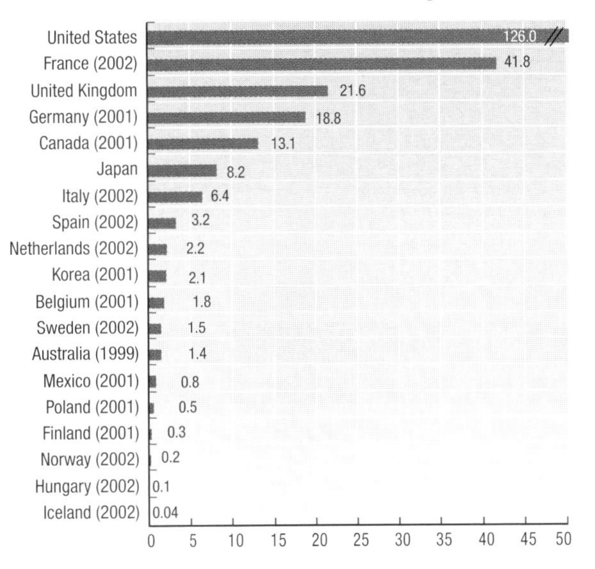

Source: OECD (2007), Structural Analysis Statistics, STAN Industry database, April.

StatLink http://dx.doi.org/10.1787/104883872725

Figure 1.1b. Total aerospace production breakdown in OECD countries, 1980, 1990, 2000 and 2001

Billions of current US dollars using PPP

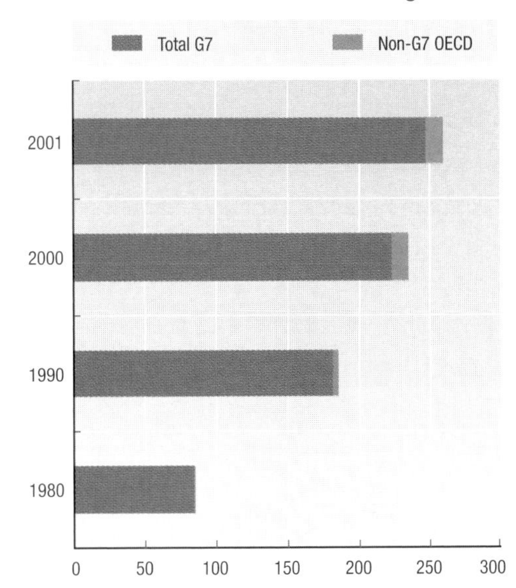

Source: OECD (2007), Structural Analysis Statistics, STAN Industry database, April.

StatLink http://dx.doi.org/10.1787/105012515873

Figure 1.1c. Breakdown of G7 aerospace industry production by year

Percentage of total aerospace production in G7 countries

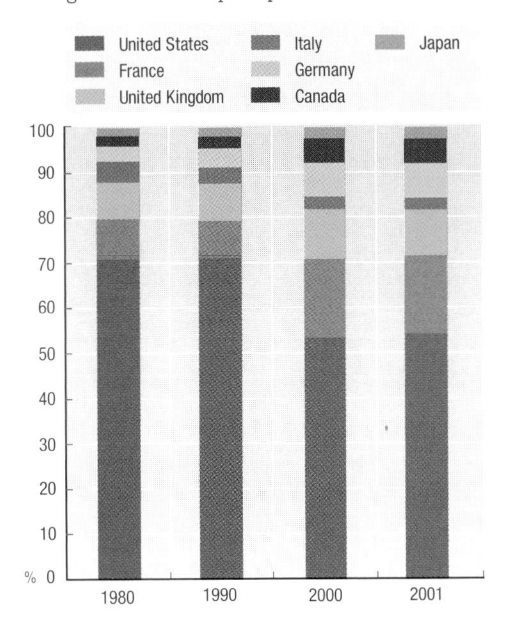

Source: OECD (2007), Structural Analysis Statistics, STAN Industry database, April.

StatLink http://dx.doi.org/10.1787/105032652352

Figure 1.1d. Average annual change in aerospace production, 1991-2001

Average percentage change in production

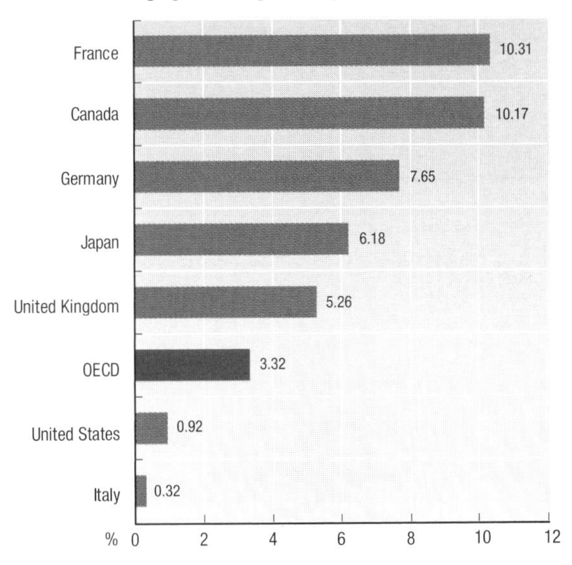

Source: OECD (2007), Structural Analysis Statistics, STAN Industry database, April.

StatLink http://dx.doi.org/10.1787/105044624671

1.1. SIZE AND GROWTH OF THE AEROSPACE SECTOR – PRODUCTION

Box 1.1. **The United Nations International Standard Industrial Classification (ISIC) Revision 3.1 Detailed Structure of Class 3530: Manufacture of aircraft and spacecraft**

This class includes:

- Manufacture of aeroplanes for the transport of goods or passengers, for use by the defence forces, for sport or other purposes.
- Manufacture of helicopters.
- Manufacture of gliders, hang-gliders.
- Manufacture of dirigibles and balloons.
- Manufacture of spacecraft and spacecraft launch vehicles, satellites, planetary probes, orbital stations, shuttles.
- Manufacture of parts and accessories of the aircraft of this class, including:
  - major assemblies such as fuselages, wings, doors, control surfaces, landing gear, fuel tanks, nacelles.
  - airscrews, helicopter rotors and propelled rotor blades.
  - motors and engines of a kind typically found on aircraft.
  - parts of turbojets and turbo propellers for aircraft.
- Manufacture of aircraft launching gear, deck arresters, etc.
- Manufacture of ground flying trainers.

This class also includes: Maintenance, repair and alteration of aircraft or aircraft engines.

This class excludes:

- Manufacture of parachutes.
- Manufacture of military ballistic missiles.
- Manufacture of ignition parts and other electrical parts for internal combustion engines.
- Manufacture of aircraft instrumentation and aeronautical instruments.
- Manufacture of air navigation systems.

Source: United Nations (2006), International Standard Industrial Classification (ISIC) Revision 3.1.

Notes: As mentioned in the Introduction, ISIC Revision 4, due in 2007, is not expected to change the aerospace classification from Revision 3.1.
This box contains a summary of what Class 3530 covers. For a comprehensive overview of what is included in the aerospace industry please refer to the United Nations *Classification Registry's* detailed structure and explanatory notes on ISIC Revision 3.1 *Class 3530: Manufacture of aircraft and spacecraft.*

THE SPACE ECONOMY AT A GLANCE – ISBN 978-92-64-03109-8 – © OECD 2007

Table 1.1. **State of the aerospace sector in 2005-2006 in selected countries**

Estimates based on national currencies and current years[1]

USA	The US aerospace industry had a successful 2006, with total deliveries projected (in late 2006) to surpass USD 184 billion, up more than 8% from USD 170 billion in 2005. While sales increased in nearly all product and customer categories, there was a 21% surge in the civil aircraft sector, with overall exports rising to USD 82 billion. Combined with relatively flat imports of aerospace products, the net trade surplus for the sector was expected to surpass USD 52 billion. There were 630 000 aerospace workers in 2006 compared with 1.1 million in 1990 (US Aerospace Industries Association, AIA, December 2006. The AIA includes more than 100 major American aerospace and defence companies, and 175 associate member companies.)
Europe	The European aerospace sector continued its upward trend, reporting EUR 121 billion in turnover in 2006 (up 7.17 % from 2005), with employment rising to 638 000 people (614 000 in 2005). (The AeroSpace and Defence Industries Association, ASD, June 2007. The ASD represents the aeronautics, space, defence and security industries in Europe. ASD members are 31 National Trade Associations in 20 countries across Europe, representing over 2 000 aeronautics, space and defence companies.)
France	French aerospace's consolidated revenues continued to grow, with 2006 revenues at EUR 25.7 billion (73% from exports). Unconsolidated revenues were up 9% compared with 2005, reaching EUR 32.1 billion. The civil sector (mainly aircraft) generated 67% of revenues. The total French aerospace workforce was 132 000 people (Groupement des Industries Françaises Aéronautiques et Spatiales, GIFAS, April 2007. GIFAS has 250 members, ranging from major prime contractors and system suppliers to small specialist companies. They cover the full spectrum of skills from the design, development and production of aerospace systems and equipment to maintenance and operation. Activities range from civil and military aircraft and helicopters to engines, missiles and armament, satellites and launch vehicles, plus major aerospace, defence and security systems, equipment, subassemblies and associated software.)
United Kingdom	The UK aerospace industry had a turnover of GBP 20 billion (63% from exports) in 2006, up by 5.5% versus 2005. New orders increased by 6% to GBP 26.2 billion. Direct employment remained at 124 234 jobs, with the sector supporting a total of 276 000 jobs in the UK, while British aerospace companies employed 48 785 people abroad and generated GBP 7.9 billion of turnover outside the UK (The Society of British Aerospace Companies, June 2007. SBAC) is the UK's national trade association representing over 2 600 companies supplying civil air transport, aerospace defence, homeland security and space products and services.)
Italy	The Italian aerospace industry generated an overall turnover of around EUR 11 billion in 2006, employing over 50 000 workers (The Associazione delle Industrie per l'Aerospazio i Sistemi e la Difesa, 2007. AIAD has more than 100 members.)
Canada	Despite the appreciation of the Canadian dollar (CAD) relative to the US dollar in 2005, the Canadian aerospace industry posted revenues of CAD 21.8 billion, virtually unchanged from the previous year. Export sales in 2005 generated CAD 18.5 billion (85% of total industry revenues) while domestic sales totalled CAD 3.3 billion, with the US remaining the single most important market for Canadian aerospace goods and services. Direct industry employment in 2005 was 75 000, up marginally from 2004 (Aerospace Industries Association of Canada, AIAC, July 2006. The AIAC represents more than 400 companies in Canada's aerospace manufacturing and services sector.)

1. Note that data presented here come from private sources (aerospace industry associations) to illustrate recent trends nationally and regionally. As such, and due to industry associations' distinct methods in data definition, collection and analysis, as well as reporting in national currencies, international comparability is very limited.

1.2. SIZE AND GROWTH OF THE AEROSPACE SECTOR – VALUE ADDED

Value added for an industry refers to its contribution to national Gross Domestic Product (GDP). It is often considered a better measure of output than basic production since it reduces the likelihood of double counting that is possible with the production approach.

Highlights

G7 data for 1980, 1990, 2000, 2001 and 2002 reveal that the total value-added output of the aerospace industry continued to rise in current US dollar terms, except in of 2002, which was slightly lower than 2001. This was also true for individual countries, except the US (which had lower output in 2000 than in 1990), and Italy and Canada (both saw a marginal decline in 2001 from 2000) (Figure 1.2a).

An examination of "aerospace value added" as a percentage of "total manufacturing value added" for the G7 shows variations depending on country and time period (see Figure 1.2b). However, overall, aerospace represents a small percentage of the total manufacturing value added. For example, in Canada and France, the percentage was just over 3% of the total in 2002. In the early 2000s, most countries saw a significant decline compared with the 1980s and 1990s. Nevertheless, the cyclical nature of aerospace activities and more recent trends show a rebound for the sector in the latest years (see also Section 1.1).

Definition

The data refer to Class 3530 of the UN International Standard Industrial Classification (ISIC) Revision 3.1, which covers the manufacture of aerospace (*i.e.* aircraft and spacecraft). Value added comprises such elements as labour costs, consumption of fixed capital, indirect taxes less subsidies, net operating surplus and mixed income. However, the exact calculation can vary depending upon the country and the extent to which taxes and subsidies are included. For example, Canada uses factor cost in valuation, while the United States uses market prices, and many other countries use basic prices.

Methodology

Value added for a particular sector or industry is residually calculated as the difference between gross production and intermediate inputs (*e.g.* energy, materials, labour and services) that are used during the accounting period. Production includes the cost of the intermediate inputs (such as previously made materials) in the value of the final product, and thus can lead to double counting (as the value of the intermediate input can also be included in the final production figures of another establishment in the same industry). The data on value added are submitted to the OECD from member countries and are included in the OECD Structural Analysis Statistics, STAN Industry database. The very limited nature of value added data for OECD countries in aerospace has circumscribed the analysis to the G7 countries.

Data comparability

Although the United Nations System of National Accounts 1993 (SNA93), along with the European System of Accounts 1995 (ESA95), requires a submission to be provided at basic prices (the value of production includes taxes less subsidies), there are some country variations (*e.g.* Canada has values at factor costs). However, attempts are made to standardise the data as much as possible.

The availability of value added data for the aerospace industry can be quite difficult to obtain. For example, 12 of the 23 non-G7 countries had no data or zero values for the aerospace industry value added. Hence, our analysis is limited to the G7 because of the availability of data and these countries' recognised significant contribution to aerospace production.

All statistics were converted from current domestic currencies into billions of US dollars using Purchasing Price Parities (see Annex for details on PPP). Note that Canadian and German data for 2002 were estimated using 2001 values and that German data prior to 1991 only refer to Western Germany.

Data sources

- OECD (2006), OECD Structural Analysis Statistics, STAN Industry database, OECD, Paris, October.

1.2. SIZE AND GROWTH OF THE AEROSPACE SECTOR – VALUE ADDED

Figure 1.2a. **Value added by aerospace industry for G7 countries, 1980, 1999, 2000, 2001, 2002[1]**

Billions of US dollars using PPP

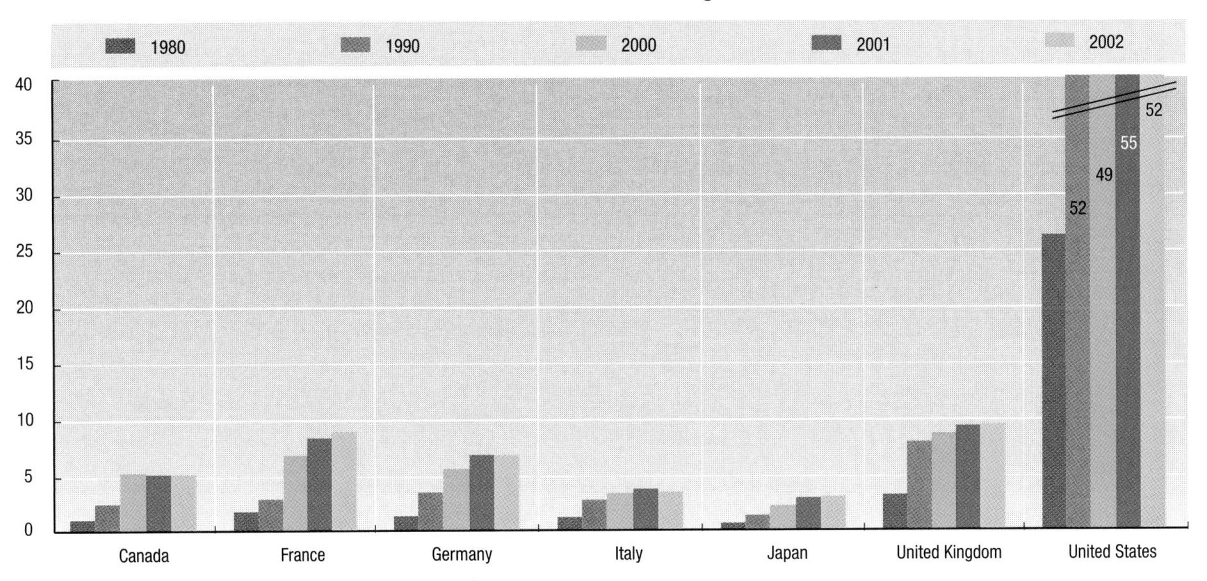

1. Canadian and German 2001 data used as 2002 estimate.

Source: OECD (2006), Structural Analysis Statistics, STAN Industry database, OECD, Paris, October.

StatLink http://dx.doi.org/10.1787/105047100357

Figure 1.2b. **Aerospace value added as percentage of national manufacturing value added for G7 countries, 1980, 1990, 2000, 2001, 2002[1]**

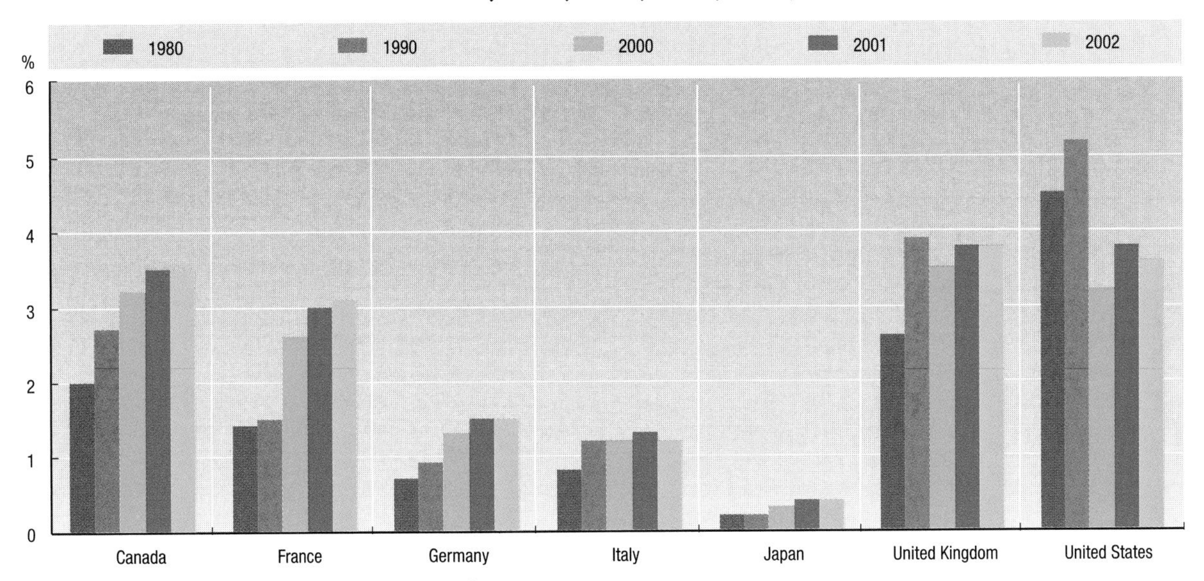

1. Canadian and German 2001 data used as 2002 estimate.

Source: OECD (2006), Structural Analysis Statistics, STAN Industry database, OECD, Paris, October.

StatLink http://dx.doi.org/10.1787/105056476328

1.3. AEROSPACE INDUSTRY RESEARCH AND DEVELOPMENT

The official OECD statistics relating to aerospace industry R&D presented here focus on business enterprise research and development (BERD) data. BERD is considered to be closely linked to the development of new products and production techniques.

Highlights

BERD data for aerospace, which totalled USD 19.8 billion in 2002, are heavily dominated by a few large countries (Figure 1.3a). Four of the OECD's largest industrial spenders - the US, France, the UK and Germany – account for 84% of the total.

Examination of BERD as a percentage of all manufacturing R&D for 1991, 1996, and 2002 shows a general decline in all G7 countries (except the UK and Japan) (Figure 1.3b). The largest proportional decline was in the US, where it fell from 18% to less than half than amount over this period. Among the non-G7 countries, the results were mixed, with some increasing and others decreasing the proportion of aerospace R&D expenditure. Furthermore, only two OECD member countries spent proportionately more on aerospace R&D in 2002 than 1991 (Spain and Norway).

Looking at expenditure for 1991, 1996 and 2002 (all expressed in US dollars using Purchasing Power Parities), the largest spender remained the US, although its expenditure declined substantially in absolute terms (Figure 1.3c). As mentioned before, the cyclical nature of aerospace activities and recent industrial trends have lead to a rebound in the US since 2002.

Definition

The aerospace industry encompasses the manufacture of a wide range of aircraft and spacecraft products (including passenger and military aeroplanes, helicopters, and gliders, as well as spacecraft, launch vehicles, satellites, and other space-related items).

Business enterprise research and development (BERD) expenditure refers to all R&D that is carried out in the business enterprise sector (*i.e.* by market-oriented firms and institutes) regardless of the source of funding. The examination here was of R&D by the aerospace industry.

Methodology

The data on R&D expenditures by the aerospace industry are based on official statistics provided to the OECD by its member countries. These data are then adjusted by the OECD for any existing deficiencies and anomalies to ensure they are comparable and consistent with OECD requirements. These data are then included in the OECD Analytical Business Enterprise Research and Development (ANBERD) database. ANBERD is subject to revisions because it depends upon a number of estimation techniques that are constantly being refined and reviewed. There are major issues when adding government R&D (GBAORD) to the business R&D presented here. This process may lead to some double counting since any government R&D funding sent to a private enterprise may appear as R&D expenditure in both accounts.

Data comparability

All statistics were originally current domestic expenditures that were converted into US dollars using Purchasing Power Parities (see Annex for details on PPP). Thus, comparability across currencies should not be a major concern. Furthermore, as the data are adjusted for deficiencies and anomalies that may have existed in the original data submitted by official respondents, the level of international comparability can be expected to be quite acceptable. The ANBERD database includes statistics on only 19 of the 30 OECD member countries.[1]

Notes

1. These 19 countries are: Australia, Belgium, Canada, Czech Republic, Denmark, Finland, France, Germany, Ireland, Italy, Japan, Korea, Netherlands, Norway, Poland, Spain, Sweden, United Kingdom, United States.

Data sources

- OECD (2006), OECD Structural Analysis Statistics, STAN Industry database, OECD, Paris, October.
- OECD (2006), OECD Analytical Business Enterprise Research and Development (ANBERD) database, OECD, Paris, September.

Figure 1.3a. **R&D expenditures in aerospace industry by OECD country, 2002**

Millions of US dollars using PPP and percentage of OECD aerospace R&D total

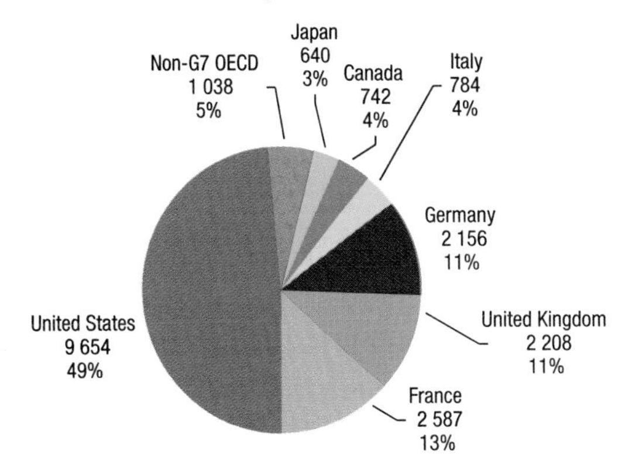

Source: OECD (2006), *Structural Analysis Statistics, STAN R&D database*, OECD, Paris, October.

StatLink http://dx.doi.org/10.1787/105065142286

Figure 1.3b. **Aerospace R&D as per cent of manufacturing R&D for selected OECD countries, 1991, 1996, 2002**

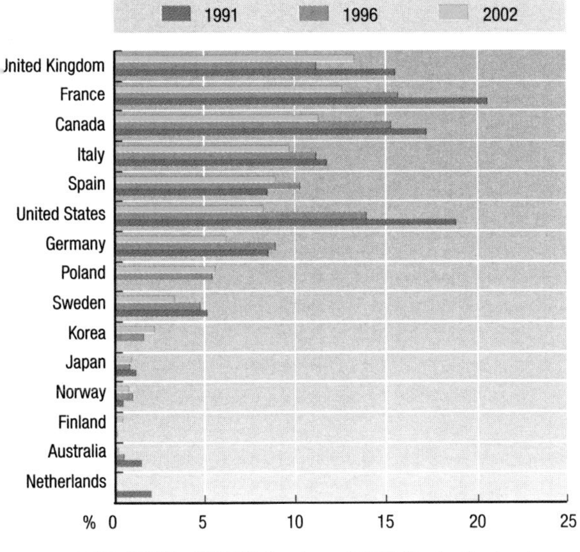

Source: OECD (2006), *ANBERD database*, OECD, Paris, September.

StatLink http://dx.doi.org/10.1787/105080617625

Figure 1.3c. **BERD of aerospace industry for available OECD countries, 1991, 1996, 2002**

Millions of current US dollars using PPP

Source: OECD (2006), Structural Analysis Statistics: STAN R&D database, OECD, Paris, September

StatLink http://dx.doi.org/10.1787/105110638536

2. READINESS: INPUTS TO THE SPACE ECONOMY

This chapter examines the technical, commercial and financial infrastructures necessary to engage in significant space activities. The focus is on the financial and human resources employed in the production of space-related hardware; the provision of related services; as well as research and development activities that may lead to the creation or improvement of goods and services.

The sections below provide an overview of government budgets for space activities (both for public space programmes and for R&D activities); the capital stock of space assets; and human capital.

Data for these areas are quite limited, and comparable official statistics are lacking. When official OECD data are perceived to be lacking in quality or quantity, or are non-existent, then non-OECD data – from both official governmental and non-official sources (e.g. industry associations and consulting groups) – are used.

2.1. Budgets for space activities

National and other institutional budgets often contribute to the start-up and development of capital-intensive and high technology sectors such as space. This section provides details on two aspects of government budgets dedicated to space activities: (1) public institutional space budgets; and (2) public space R&D budgets. A number of methodological caveats are necessary. First, it is important to note that institutional budgets generally cover annually allocated amounts; however, the actual expenditures may vary from these estimates. Second, while every effort has been made to make the data as comparable as possible, the data on institutional budgets come from a variety of sources and, hence, pose data comparability challenges. Third, some data for R&D may not be included because the data are not publicly available for reasons of national or corporate security.

2.1.1. PUBLIC INSTITUTIONAL SPACE BUDGETS

As the number of countries with space programmes continues to rise, so do government space budgets for military and civilian applications. This section examines the size and areas of funding of space programmes of selected OECD and major non-OECD countries.

Highlights

Over thirty countries have dedicated space programmes and more than fifty have procured satellites in orbit, mostly for communications purposes (Figure 2.1.1b).

In 2005, civilian and military budgets for space programmes of OECD countries totalled about USD 45 billion (although data were not available for some smaller OECD countries). Of this amount, over 81% was accounted for by the US, followed by France, Japan, Germany and Italy (Figure 2.1.1c). A breakdown of other countries reveals relatively large contributions by countries such as Sweden and Switzerland (Figures 2.1.1e and 2.1.1f).

US space budgets picked up, especially after 2001, with the 2006 estimated budget more than 30% higher than five years earlier (see Figure 2.1.1d). The general trend shows US military space budgets (i.e. Department of Defense) rising as a percentage of the total, especially since 2001 (see Figure 2.1.1g).

Europe also budgeted significant amounts for space programmes: about USD 6 billion in 2005. An examination of European budgets shows that the three largest contributors (France, Germany and Italy) account for 76% of the overall European total, including 90% of national and 68% of ESA budget totals (see Figure 2.1.1h).

Several non-OECD countries have also significantly boosted their civilian space investments over the past few years and are continuing to do so. In 2005, the Russian space budget was estimated at USD 647 million (18.3 billion roubles) and India's budget at USD 714 million (31.48 billion rupees), with Brazil's at USD 92 million (223 million reals). The Chinese budget was estimated at a tenth of NASA budget by Chinese officials, or around USD 1.5 billion in 2005.

The national public space budget as a percentage of GDP for 2005 was largest in the US, at 0.295%, about three times higher than France (Figure 2.1.1h). The top ten OECD countries included all the G7, except the UK. Also, note that the three major non-OECD space countries, India, Russia, and China, all ranked within the top five, ranging around the G7 average of 0.084%.

Definition

Space budgets refer to the amounts that governments have indicated they will provide to public sector agencies or organisations to achieve space-related goals (*e.g.* space exploration, better communications security). For OECD countries examined here, the space budgets may serve both civilian and military objectives. However, significant portions of military-related space budgets may not be revealed in published figures. Data for non-OECD countries Brazil, Russia and India refer to civilian and/or dual-use programmes. Chinese figures are only estimates and not official data. Other estimates of China's space budget (from diverse Western and Asian sources) range from USD 1.2 to more than USD 2 billion.

Methodology

Estimates were done using primarily official documents, but also private data (see sources). All figures were converted into 2005 US dollars, using market exchange rates from the OECD National Accounts databases. All values are in current US dollars, and adjustments for inflation have not been made.

Looking at public budgets related to space poses several methodological challenges. First, when they are available publicly in some detail, budgets may not necessarily match current expenditures. Second, published budgets may not reveal certain confidential segments of space programmes (*e.g.* for military purposes). Third, some expenditure may be classified under other areas of government expenditure, *e.g.* telecommunications or R&D, and not under "space". Finally, data were not available for all OECD member countries (although they were available for all major space participants). Data were not available for Australia, Iceland, Mexico, New Zealand, the Slovak Republic and Turkey.

Concerning the number of operational satellites by country, mean estimates have been derived from different figures found in the existing literature and databases. Data presented here should provide reasonable orders of magnitude.

Data comparability

Although the data presented here provide a first impression of space budgets, comparing the size of budgets across countries raises a number of issues, such as taking into account differences in budget lines (different definitions across countries), currencies and Purchasing Power Parities (see Annex for details on PPP). As an example, expenditures in currently low per capita income countries, such as China and India, may have a higher purchasing power (e.g. because labour and services cost less) than similar expenditures in high per capita income countries. Thus, the real, PPP-adjusted expenditure in such countries may be higher than indicated by a comparison based solely on exchange rates.

Data sources

- Aerospace Industries Association (2006), *US Aerospace Facts and Figures 2005-2006*, Washington DC.
- Brazil Ministro da Ciência e Tecnologia (2006), *Agência Espacial tem a melhor execução orçamentária dos últimos anos (Space agency has better budget execution than in past years)*, 23 February.
- Canadian Space Agency (2006), *Annual Report*, Montreal, Canada.
- European Space Technology Platform (ESTP) (2006), *Strategic Research Agenda*, V1.0, June.
- Indian Ministry of Finance (2006), *Notes on Demands for Grants, 2006-2007*, No.88, Department of Space.
- JAXA (2006), *Annual Report*, Tokyo, Japan.
- NASA (2006), *Aeronautics and Space Report of the President, Fiscal Year 2005 Activities*, Washington DC.
- OECD (2006), *Annual National Accounts, Exchange Rate, PPPs, and Population database*, OECD, Paris, France.
- OECD (2007), *National Accounts of OECD Countries, Volume I - Main Aggregates*, OECD, Paris, France.
- OECD/IFP research (2007), *various satellite database sources* (*e.g.* Eurospace, *European Space Platform Database*, January 2007; UCS, *Satellite Database*, December 2006), February.
- Roscosmos (2006), *The Federal Space Program of Russia*, Website: http://www.roscosmos.ru (in Russian), Moscow, Russia.
- Xinhua News Agency (2006), *Chinese Annual Space Budget to Exceed Two Billion Dollars*, October 12.

Figure 2.1.1a. **Public space budgets as a per cent of national GDP for available OECD and non-OECD[1] countries, 2005**

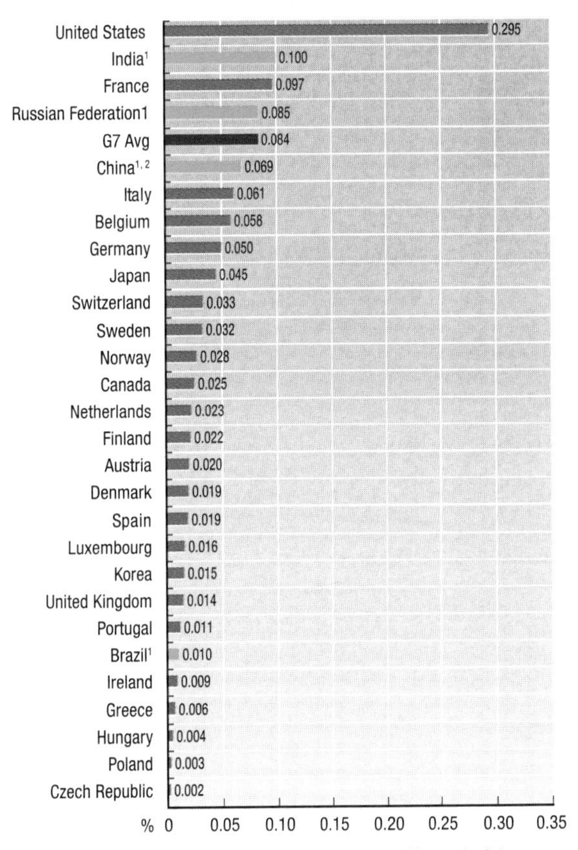

1. Non-OECD countries are Brazil, Russia, India and China.
2. Chinese data based on unofficial estimates.

Sources: Budgets: NASA, CSA, ESTP (Europe), JAXA, other national sources.
GDP: OECD (2007), *National Accounts of OECD Countries, Volume I – Main Aggregates*, OECD, Paris, France.

2.1.1. PUBLIC INSTITUTIONAL SPACE BUDGETS

Figure 2.1.1b. **Countries with operational satellites in orbit as of December 2006 (estimates)**[1]

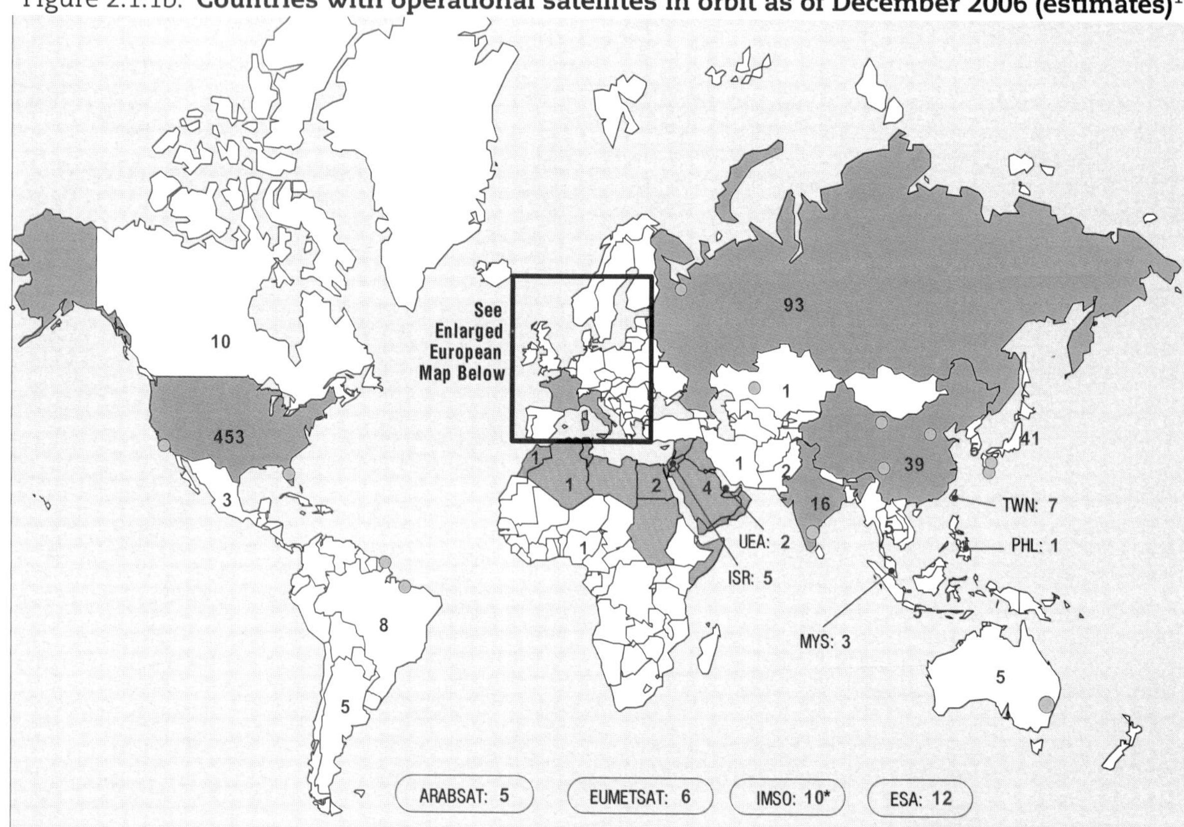

Countries with space budgets over 0.06% as a % of GDP

Members of the European Space Agency, ESA (17 European countries + Canada)

European Space Agency's Cooperating States (4 PECS countries)

Members of the Arabsat satellite communications organisation (21 countries)

Operational satellites in international organisations

1 Number of operational civilian/dual-use satellites as of Dec. 2006*

Selected major space launch sites

* NOTE: The estimated number of operational satellites in orbit covers governmental and commercial telecommunications, earth observation, scientific and dual-use satellites (including military when known), but not exploration probes (*e.g.* Venus Express). Some international joint missions are accounted for under a lead country, as to avoid double-counting, and several countries' data include rather large commercial fleets of telecommunications satellites (USA with Intelsat; France with Eutelsat; Luxembourg with SES Global). Data were extracted from several databases of satellites, referenced in the text (under "Data sources"), and the figures represent orders of magnitude, not official estimates.

* The International Mobile Satellite Organisation (IMSO) is the intergovernmental organisation (88 member states) that oversees certain public satellite communication services (*e.g.* maritime search and rescue co-ordinating communications) provided via the satellites of the Inmarsat Ltd company.

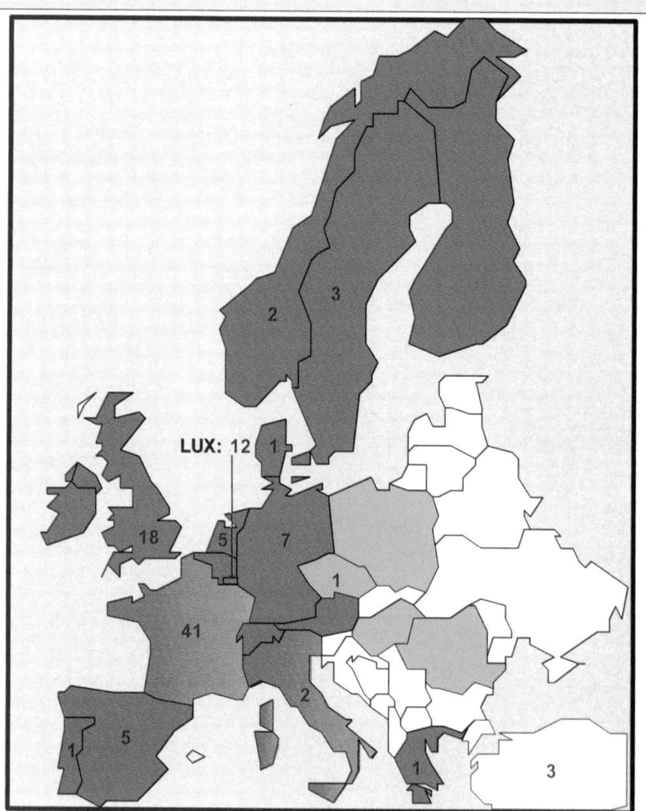

Source: OECD/IFP research (2007).

StatLink http://dx.doi.org/10.1787/105150204463

Figure 2.1.1c. **Space budgets of selected OECD and non-OECD[1] Countries, 2005**

Billions of current US dollars

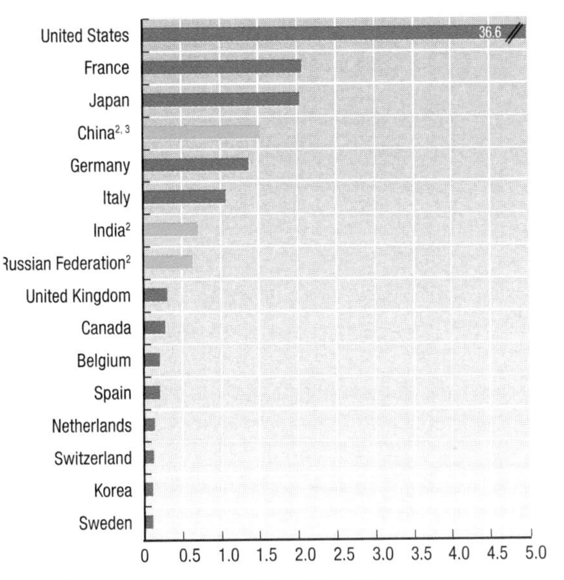

1. For budgets greater than USD 100 million.
2. Non-OECD countries are Russia, India and China.
3. Chinese data based on unofficial estimates.

Sources: Budgets: NASA, CSA, ESTP (Europe), JAXA; other national sources, 2006.

GDP: OECD (2007), *National Accounts of OECD Countries, Volume I – Main Aggregates*, OECD, Paris, France.

StatLink http://dx.doi.org/10.1787/105155725065

Figure 2.1.1d. **US government total space budget, 1990-2007[1]**

Billions of current US dollars

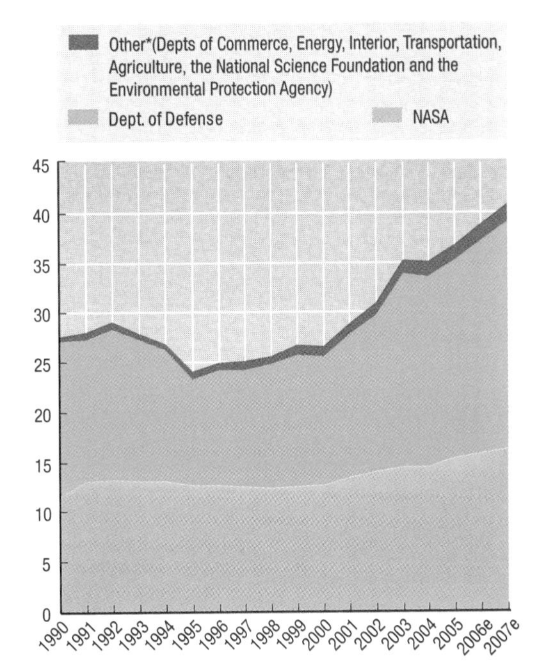

1. 2006 and 2007 data are estimates.

Source: NASA (2006), *Aeronautics and Space Report of the President, Fiscal Year 2005 Activities.*

StatLink http://dx.doi.org/10.1787/105180638342

Figure 2.1.1e. **Breakdown of total space budgets for OECD countries, 2005**

Country space budget as % of total OECD budget for space programmes[1]

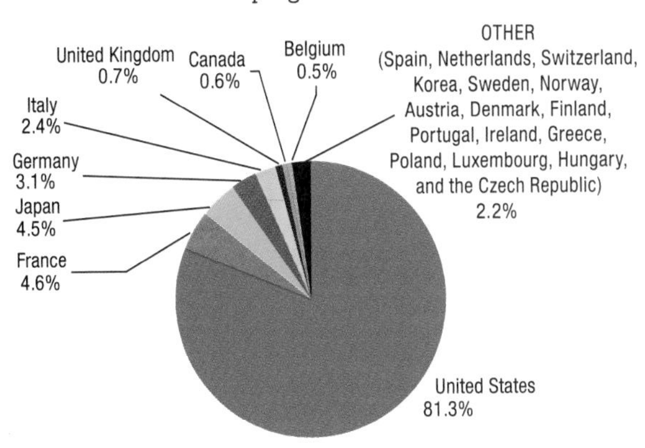

1. Data not available for Australia, Iceland, Mexico, New Zealand, the Slovak Republic and Turkey.

Sources: Budgets: NASA, CSA, ESTP (Europe), JAXA; other National sources, 2006

Figure 2.1.1f. **Breakdown of other OECD space budgets, 2005**

Millions of current US dollars (Total= USD 1,021 Million)

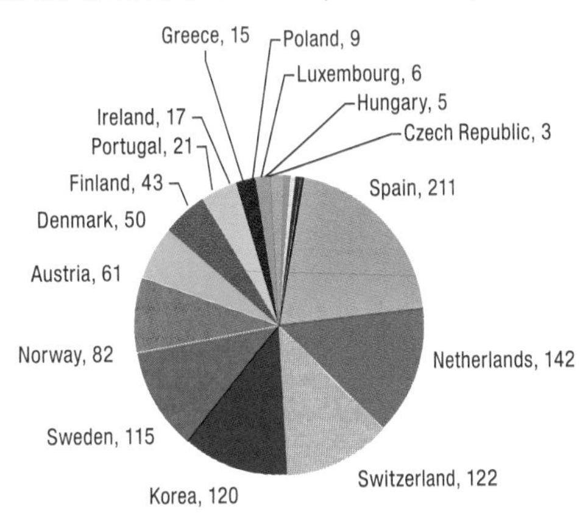

Source: European Space Technology Platform (ESTP) (2006), *Strategic Research Agenda*, Version 1.0, June.

2.1.1. PUBLIC INSTITUTIONAL SPACE BUDGETS

Figure 2.1.1g. **Military[1] as per cent of US total space budget ,1990-2007[2]**

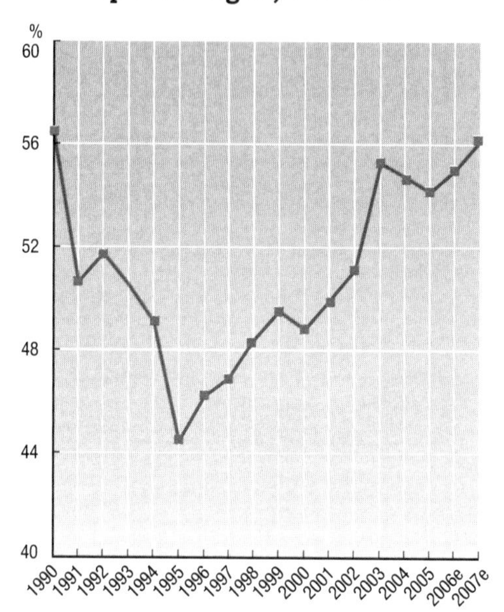

1. Military is Department of Defense.
2. 2006 and 2007 data are estimates.

Source: NASA (2006), *Aeronautics and Space Report of the President, Fiscal Year 2005 Activities.*

Figure 2.1.1h. **Breakdown of selected European space budgets[1] for space, 2005**

Millions of current US dollars

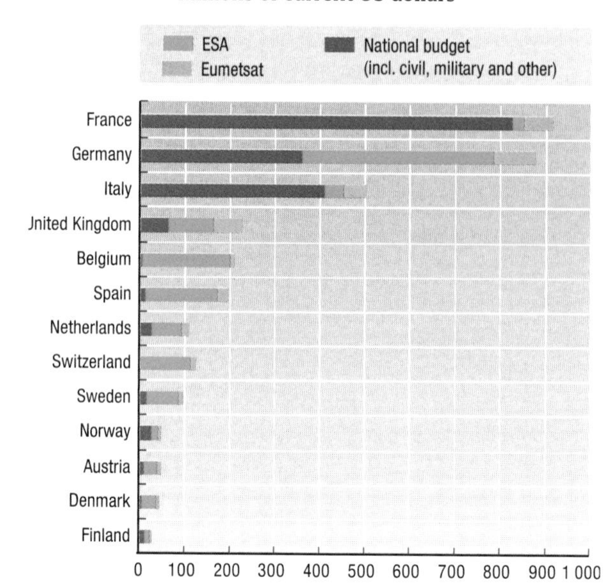

1. Only those countries with budget of greater than USD 25 million.

Source: European Space Technology Platform (ESTP) (2006), *Strategic Research Agenda*, Version 1.0, June.

2.1.2. PUBLIC SPACE RESEARCH AND DEVELOPMENT BUDGETS

Since the beginning of the space age, government support for research and development (R&D) in the space sector has been crucial for developing civilian systems and applications. To complement the general information provided by institutional space budgets (section 2.1.1), an analysis of GBAORD (Government Budget Appropriations or Outlays for R&D) is presented.

Highlights

According to GBAORD data, total government budgets of OECD countries for space-related R&D were USD 16.4 billion in 2004 (Figure 2.1.2a) with a few large countries dominating the total. An examination of GBAORD data for OECD and selected non-OECD countries reveals that the G7 dominated many of the top positions, with the US leading with a budget of USD 10.6 billion (Figure 2.1.2b). Data from non member countries were included to put these figures into perspective.

However, an examination of government space R&D as a percentage of total civilian GBAORD for the OECD area reveals that some non G7 members (e.g. Spain, Netherlands) allocated as much as G7 countries (Figure 2.1.2c). A more comprehensive examination of space R&D as a percentage of total civilian GBAORD from 1981 to 2005 shows that the US government allocated more as a percentage of all civilian R&D to space than all other major players (Figure 2.2). Other countries with relatively high space R&D expenditures include France and Belgium. Most countries also appear to follow a similar pattern with expenditures peaking in the early to mid 1990s and decreasing at least until 2005.

Definition

GBAORD data are assembled by national authorities analysing their budget for R&D content and classifying these R&D outlays by "socio-economic objective", which can include both civilian and military outcomes. These objectives represent the intention of the government at the time of funding commitment. The GBAORD data here are those included under the socio-economic objective of "exploration and exploitation of space" for civilian purposes only. Government-funded R&D may be carried out domestically or abroad by business enterprises, governments, higher education institutions or the non-profit sector.

Methodology

Government support for R&D in civilian space programmes can be measured using two methods. One is to examine all government units and add up their total R&D expenditures related to civilian space activities. The other is to examine how much governments officially state that they will spend on various socio-economic objectives (e.g. exploration and exploitation of space). This latter method uses GBAORD.

One caveat to this approach is that GBAORD only indicates what governments state they will spend in their budget; the actual expenditure may be different from the allocated total. Also, governments tend to link R&D expenditures using GBAORD to their "primary" socio-economic objective. As such, if space is a secondary or tertiary objective of R&D expenditure, then it might not be reflected in the totals. In addition, part of the budget allocated to space may also fall under military or defence-related R&D, which is not considered here but can be substantial in some countries, as seen in Section 2.1.1. Finally, it is possible for some countries to break down the socio-economic objective of "space exploration and exploitation" into further sub-objectives. Although this is suggested by the European Statistical Office (Eurostat), it is not presently required by the OECD, and very few countries are currently able to provide this breakdown.

Data comparability

Data comparability may be affected by the fact that GBAORD tends to represent expenditures of the federal or central government only. The OECD Frascati manual, which provides useful guidelines for R&D comparisons, does suggest the inclusion of provincial/state data if they are "significant". Thus, comparability may be limited to the extent that data compilers perceive expenditures of other levels of government as significant. Also, several countries with considerable space programmes are not included, due to current lack of GBAORD data (*e.g.* Brazil, China, India). And a few OECD countries lacked data for some years (*e.g.* Italy from 2002 to 2004), in which case, their first and last available year of data were used (Figures 2.1.2aand 2.1.2b). Prior to 1991, German data referred only to Western Germany. All figures were converted into US dollars using Purchase Power Parities (see Annex for details on PPP).

Data sources

- OECD Science, Technology and R&D Statistics (2006), *Main Science and Technology Indicators*, August.

2.1.2. PUBLIC SPACE RESEARCH AND DEVELOPMENT BUDGETS

Figure 2.1.2a. **Breakdown of total OECD GBAORD for space, 2004**[1]

Percentage of total OECD GBAORD for space by country

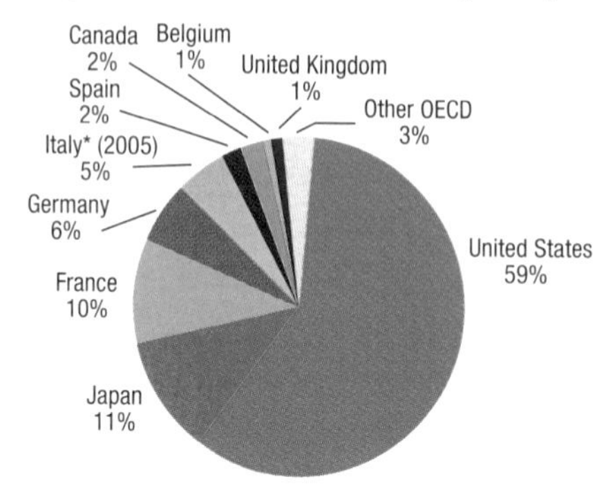

1. Data not available for Italy in 2004.

Source: OECD Science, Technology and R&D Statistics (2006), *Main Science and Technology Indicators*, August.

StatLink http://dx.doi.org/10.1787/105181570667

Figure 2.1.2b. **GBAORD for space programmes in available OECD and selected non-OECD[1] countries, (latest year)**

Millions of current US dollars using PPP

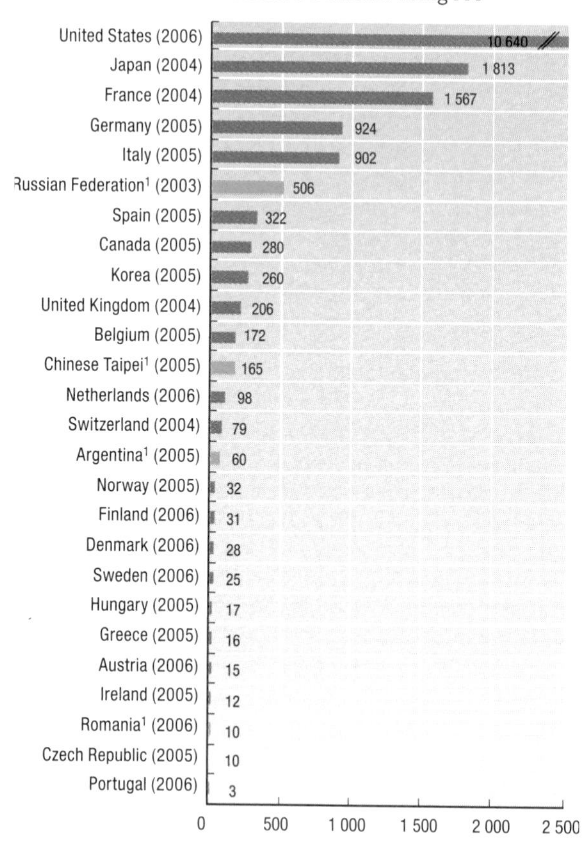

1. Non-OECD Countries: the Russian Federation, Chinese Taipei, Argentina, Romania.

Source: OECD Science, Technology and R&D Statistics (2006), *Main Science and Technology Indicators*, August.

StatLink http://dx.doi.org/10.1787/105230243001

Figure 2.1.2c. **Space as percentage of national civil GBAORD for OECD countries, 2004 (or latest year)**

Space GBAORD as per cent of total national civilian GBAORD for available OECD countries

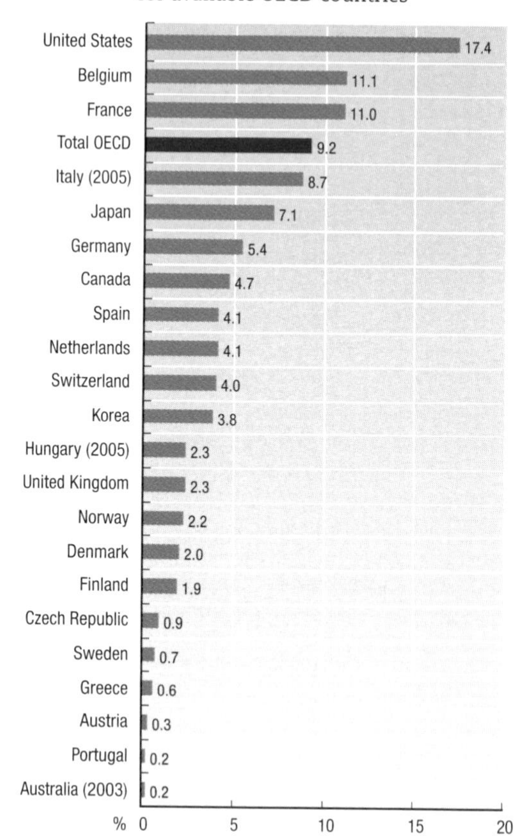

Source: OECD Science, Technology and R&D Statistics (2006), *Main Science and Technology Indicators*, August.

StatLink http://dx.doi.org/10.1787/105306014530

2.2. Space R&D as a percentage of national civilian R&D for selected OECD countries, 1981-2005[1]

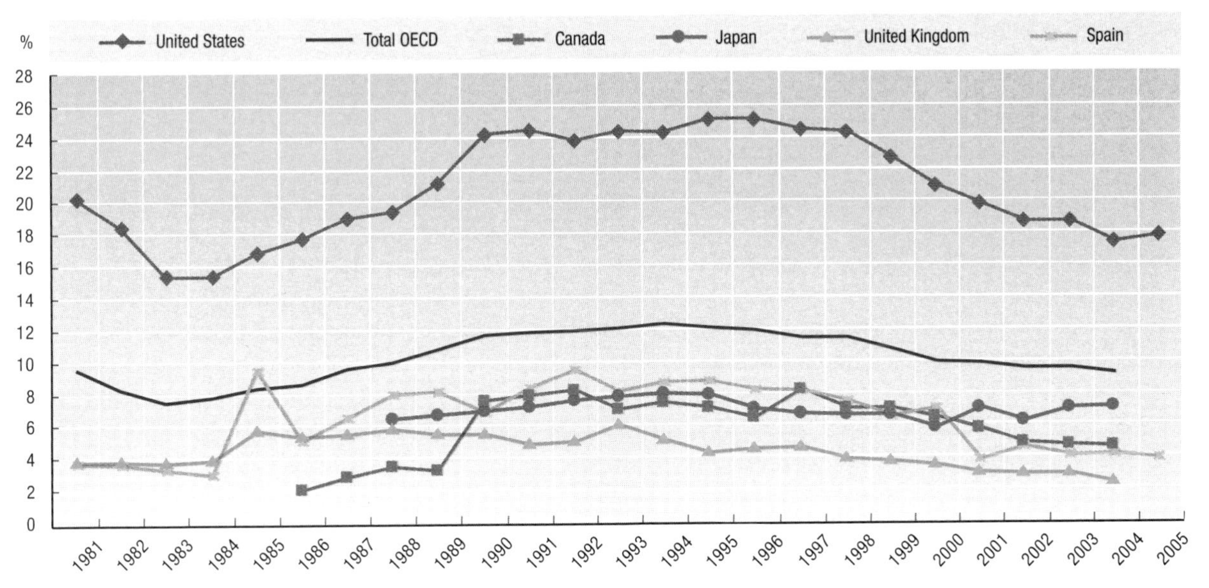

StatLink http://dx.doi.org/10.1787/105330567117

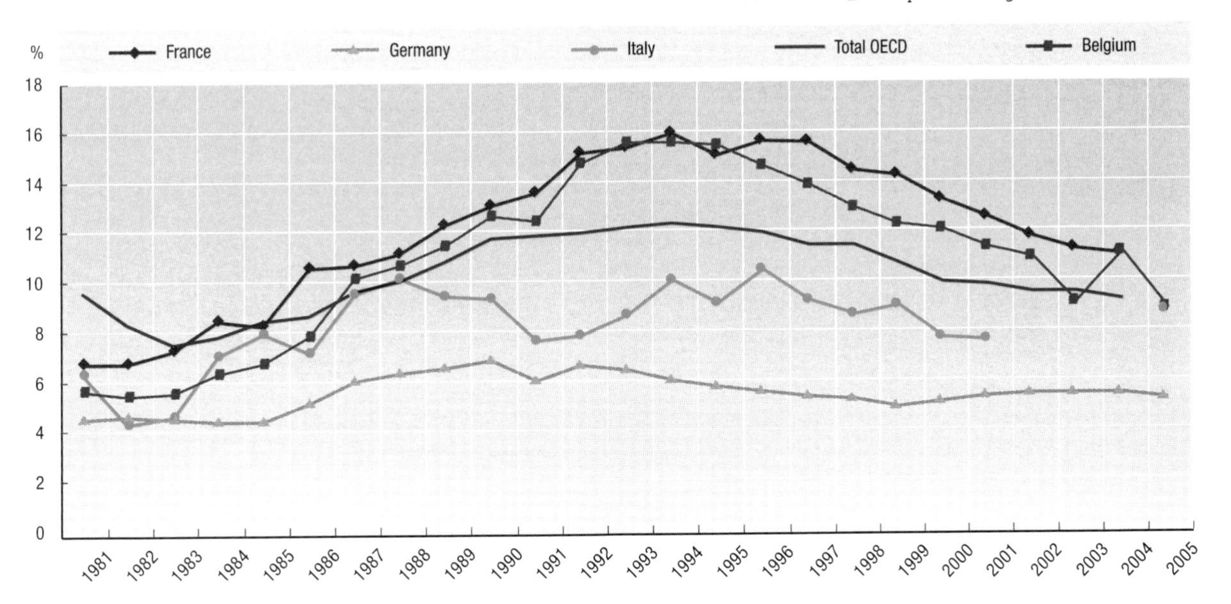

1. Where data available.

Source: OECD Science, Technology and R&D Statistics (2006), *Main Science and Technology Indicators*, August.

StatLink http://dx.doi.org/10.1787/105358126265

2.2 CAPITAL STOCKS OF SPACE ASSETS

Capital stocks represent the accumulation of equipment and structures available to produce goods or render services. In the case of space activities many of the installations are predominantly of a public nature (*e.g.* laboratories, launch pads) although the private sector has an increasingly important role in providing services. Because the budget sources are so diverse, they are difficult to estimate. However, satellites, as discrete in-orbit assets, can provide more easily measurable indications of the value of space infrastructure.

Highlights

Satellites, as in-orbit assets, have strategic as well as economic value. The estimates here provide orders of magnitude. Around 940 satellites are currently operating in orbit, with more than two-thirds being communication satellites, many in geostationary orbits. Satellites in geostationary orbit are located at an altitude of 36 000 kilometres above the equator, and as seen from the Earth remain perfectly stationary in the sky. This allows one satellite to cover large parts of the Earth, with useful applications (e.g. international communications and direct to home satellite television, meteorology). There are relatively few geostationary positions, and they are distributed to countries by the International Telecommunications Union. The development of small and more affordable satellites by current and new space-faring countries will contribute to increased traffic and space debris in busy orbits over the next decades (OECD, 2005). Man-made space debris includes diverse space objects (e.g. non-functioning satellites, parts of rockets) which – depending on their respective altitude and orbit – may take years to millennia to disintegrate in the atmosphere. Nine hundred commercial and governmental satellites were launched over 1997-2006 (65 to 110 satellites each year); and in the next decade, the number of satellites to be launched could rise to 960 (an increase of 6.6%) (Euroconsult, 2006). Over 10 000 objects (including parts of launchers, exploration probes and demonstrators) have been launched into space since 1957. The standard operational lifetime of satellites in orbit varies from a few weeks to almost twenty years.

A 2005 study estimated that the 937 satellites operating in the Earth's orbit at that time had a replacement value ranging from USD 170 to 230 billion (Odenwald, 2005). As a further illustration, a recent OECD/IFP study on water resources management (see Annex B) estimates a very conservative value of the stock of 100 active Earth observation (EO) satellites at around USD 20 billion (current). The stock value of EO satellites launched in 2006 – a particularly active year – amounted to USD 3.2 billion or around 15% of the total space-based Earth observation and meteorological infrastructure at that time (Box 2.2). This value should rise, as more satellites are launched by emerging space countries, and as a growing number of international EO missions contribute to the monitoring of global climate change.

Space-based assets have rather short operational lives, and the relative levels of investment (*i.e.* annual spend as a proportion of total assets) for satellite infrastructure are higher than those for road network infrastructures, but they are quite close to those for capital expenditures on rail networks and telecommunications (Table 2.2). In that context, and taking into account sunk costs due to research and development, the rate of replacement and expansion of the space infrastructure seems relatively low overall compared with that for terrestrial infrastructures.

Definition

Capital stocks provide an economic valuation of the equipment and the structures (*e.g.* buildings) that are available to produce goods or provide services. Operational satellites that currently orbit the Earth are used here as examples of strategic and valuable economic assets.

Methodology

The estimates provide mainly orders of magnitude. As in the case of other technology-intensive infrastructures, satellites are the visible outcomes of necessary but long-term civilian and/or military R&D investments made by public investors, and often not accounted for in the satellites' published costs. Sustained investments in scientific and technology fields are essential prerequisites for any active space-based infrastructure (*i.e.* from space launchers to in-orbit systems).

In the case of the OECD/IFP estimates of capital stocks, the majority of the identified Earth observation satellites particularly useful for water management are deployed to low earth orbits (around 80 satellites), while a smaller number (around 20) are deployed to geostationary orbits, primarily for global weather forecasting. This internal study (see Annex B) used publicly available cost estimates and a GDP deflator to calculate costs at current US dollar rates (see Box 2.2).

Data comparability

International differences are important when determining capital stocks, as purchasing power parity considerations directly impact valuation (see Annex C). As an example, the value of a Western standard satellite with remote sensing instruments may range between USD 200 and 400 million. The OECD list of Earth observation satellites includes (1) many non-Western satellites (Indian, Russian, Chinese), for which values are mainly estimates; and (2) the increasingly popular small satellites carrying fewer instruments which are often less expensive than "standard" ones (USD 20 to 100 million). In addition, many satellites are more akin to high-technology prototypes than just standard industrial equipment; hence their values are difficult to estimate accurately.

Data sources

- Odenwald S. (2005), "Forecasting the Impact of a 1859-Calibre Superstorm on Satellite Resources", *Advances in Space Research*, September.
- OECD (2005), *Space 2030: Tackling Society's Challenges*, OECD, Paris.
- OECD/IFP research (2007), various satellite database sources (*i.e.* Eurospace, European Space Platform Database, accessed January 2007; Jonathan's Space Report, *Launchlog*, December 2006; UCS, *Satellites Database*, accessed December 2006).
- Satellite Industry Association (SIA) (2006), *Annual Report 2005*, SIA, Washington DC.

Table 2.2. **Estimated annual world infrastructure expenditure (additions and renewal) for selected sectors, 2005, in USD**

Type of infrastructure	Stock (USD)	Annual investment (USD)
Road	6 trillion	220 billion
Telecoms	3.2 trillion	650 billion
Rail	630 billion	50 billion

Source: OECD (2006), Infrastructure to 2030: Telecom, Land Transport, Water and Electricity, OECD, Paris.

Box 2.2. **Methodology used to assess the present-day value of the stock of 100 Earth observation satellites (including 20 meteorology satellites) active in 2006**

To value the cost of active civilian Earth observation satellites launched between 1990 and 2006, conservative public cost estimates have been used. A total cost for each given mission was estimated, using public data (*e.g.* agency or industry reports, press releases). These do not always clearly discriminate between launch costs, satellite costs and operating costs. Therefore, the figures used remain largely conservative: (a) many satellites have been operating in orbit for longer than their designed lifetime (*i.e.* inducing higher operating costs than the ones anticipated in some of the available estimates); (b) the values calculated may underestimate the overall costs of the missions (especially when estimates are not official, as in the case of China or Russia); and (c) delays in launching or possible cost overruns in developing ground based infrastructures (for operations) may not appear in the total value of the systems. Finally, a satellite mission is often the result of R&D investments over years that may not be fully reflected in the systems' available cost estimates.

A methodological approach was then developed to transform past values of satellites into present values. Although a number of specific variables could have been used to re-calculate the past value of satellites into present values (e.g. Producer Price Index or Consumer Price Index), preference was given to the "Deflator for GDP at Market Prices for the OECD-Region". There were a number of reasons for this choice. First, there was no particular variable that seemed uniquely geared towards representing the changing value of producing, launching, and servicing satellites over this period, making GDP a good variable for general price changes in the economy. Second, many of the available variables often lacked values for key international players or for certain time periods. Third, a large number of satellite launches over this period were carried out by OECD countries. The particular calculation that was performed was an "inflation" of past values into present value terms. (Usually, the purpose of a deflator is to deflate the value of an item into the prices of an earlier base year. Here, the opposite occurred. For example, if it cost USD 100 to produce an item in 2005 and USD 110 to produce it in 2006, the 10% change in prices that would usually be used as a "deflator" to put 2006 values into 2005 values, was used as an "inflator" to increase the value of 2005 into 2006 terms. Both outputs would now be valued at USD 110 dollars.) This "inflator" method was used to move all items from 1990 up to 2006 values.

2.3. HUMAN CAPITAL

Human capital is key to the development and sustainability of the space sector. The sector is home to highly skilled professionals (*i.e.* technicians, scientists and engineers).

Highlights

Although estimates vary, the space-manufacturing sector is not a very large employer compared to other labour-intensive sectors.

The European space manufacturing industry, as defined by the Eurospace industry association, had approximately 28 000 employees in 2005, rising to 29 000 employees in 2006, recovering slightly after a downward trend since 2001 (Figure 2.3a). Productivity (measured as consolidated turnover divided by number of employees) continued to rise sharply for the third straight year. Approximately 40% of total European space manufacturing employees worked in France (Figure 2.3b). France, Germany and the United Kingdom were the leading employers in space industrial activities (*i.e.* systems integrators, subsystems suppliers and equipment suppliers) (Figure 2.3c). Data for the largest OECD player in the space sector, the United States, showed total employment of approximately 66 000 in 2005, which accounted for 0.45% of total manufacturing employment (Figure 2.3d).

Definitions

Data on space-related human capital are very fragmented. Official employment statistics on the sector are poor, lacking in both quality and detail. To some extent, the gaps can be filled by non-official statistics, mainly from industry associations. The data here focus on the space manufacturing industry, and the larger services sector is not included.

The US data include the manufacturing and development of guided missiles and space vehicles (and related parts and auxiliary equipment). The US data are based on a survey by the US Bureau of Census on manufacturers of space-related items (including guided missiles and spacecraft). In the US Census Bureau's Annual Survey of Manufactures (2005) "space industry" is defined as NAICS 336414 (Guided missiles and space vehicle manufacturing) plus 336415 (Guided missiles and space propulsion units and propulsion unit parts manufacturing) plus 336419 (Other guided missile and space vehicle parts and auxiliary equipment manufacturing).

The European data come from the Eurospace association, which collected them from its members via surveys and use of supplementary data; the results are consolidated to avoid double counting and corrected for any possible errors. The European data focus on the industrial manufacturing activities of commercially oriented manufacturing units in Europe (*i.e.* involved in the development and production activities of spacecraft, space launchers and associated professional ground segments). Employment may also be defined differently between the various European countries, and between the US and Europe.

Data comparability

As mentioned previously, with the lack of official statistics on space data, the statistics on employment in the space economy can be as specific or wide as the available data permit, with challenging comparability issues. Further space-related employment information on individual countries can be found in Chapter 5.

Data sources

- Eurospace (2006), Facts and Figures: The European Space Industry in 2005, Paris.
- Eurospace (2007), Facts and Figures: The European Space Industry in 2006, Paris.
- US Census Bureau (2005), Annual Survey of Manufacturers, Washington, DC, December.

Figure 2.3a. **European space industry productivity[1] and employment, 1992-2006**

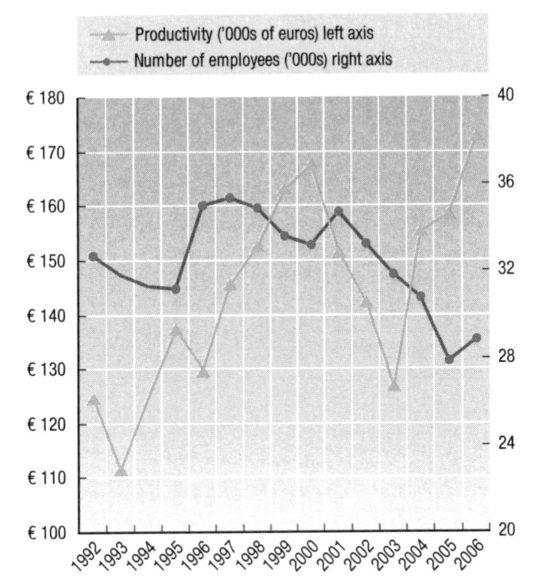

1. Productivity = thousands of euros turnover / employees

Source: Eurospace (2007), *Facts and Figures*.

Figure 2.3b. **European space industry employment by country, 2006**

Number of employees

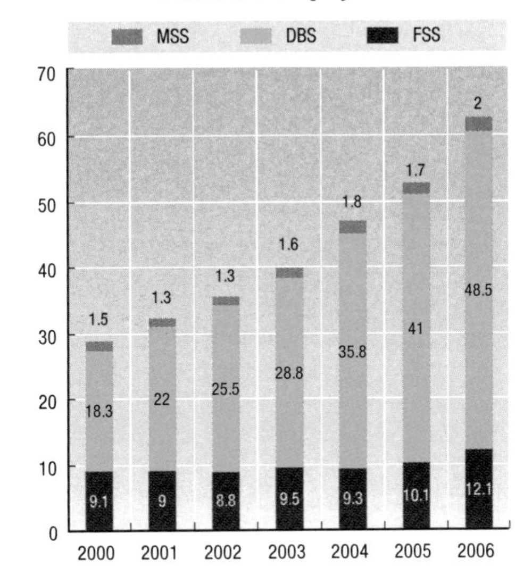

Source: Eurospace (2007), *Facts and Figures*.

Figure 2.3c. **European space industry employment by country and company type[4], 2005**

Actual number of employees

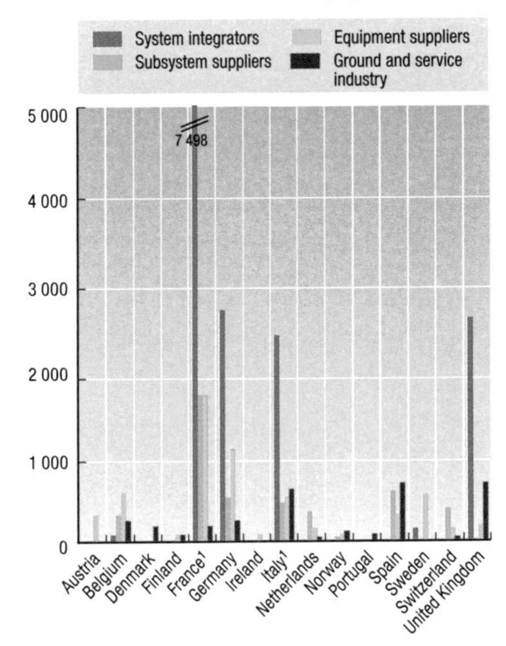

4. Data for France and Italy adjusted for non-space.

Source: Eurospace (2006), Facts and Figures.

Figure 2.3d. **US space industry[1] employment numbers and percentage of total manufacturing, 1997-2004**

Employment in thousands and % of all manufacturing employees

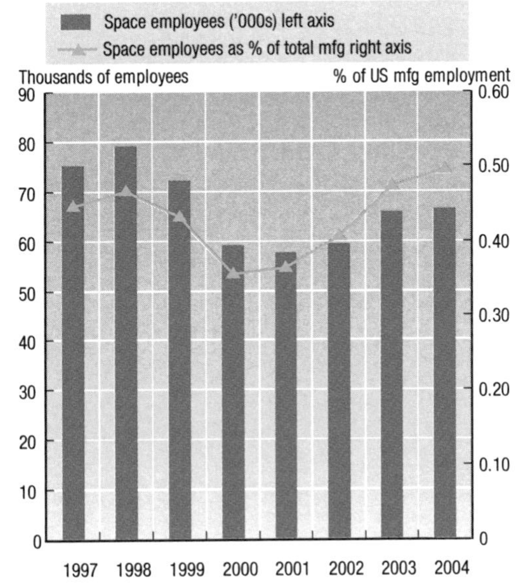

1. US Space Industry includes codes: 336414 (Guided missiles and space vehicle manufacturing) + 336415 (Guided missiles and space propulsion unit and propulsion unit parts manufacturing) + 336419 (Other guided missile and space vehicle parts and auxiliary equipment manufacturing).

Source: US Bureau of Census (2005), *Annual Survey of Manufacturers*, Washington, DC.

3. INTENSITY: OUTPUTS FROM THE SPACE ECONOMY

The statistics on outputs offer an overview of the use of space infrastructures, i.e. products or services that are produced or provided by the space sector. Outputs also include the benefits to industries or countries deriving from the production of space products or the performance of space-related R&D. These include financial benefits (e.g. trade revenues) and indicators of present and future financial benefits (e.g. patents).

The indicators examined here incorporate: (1) revenues from the space industry; (2) space-related services; (3) international trade in space products; (4) space patents; and (5) space launch activity and payloads (i.e. a satellite or an instrument on a satellite). Two of these sections – international trade in space and space patents – draw upon official OECD statistics. Statistics on revenues, services and launches examine data and information from non-OECD sources, such as governments, industry associations and consulting firms.

3.1. REVENUES FROM THE SPACE INDUSTRY

The space industry covers many segments. Using mainly private sources, the statistics presented here provide orders of magnitude for three major manufacturing segments of the satellite industry: satellite manufacturing, ground equipment and the launch industry. Space-related services are covered in the next section.

Highlights

Worldwide satellite industry revenues remained steady from 2002 to 2005 at USD 35-36 billion, with an increase in 2006 for the manufacturing segment, which attained levels similar to 2000. (Figure 3.1a). A more global recovery over time is anticipated, based on the cyclical nature of space activities (*e.g.* renewal of satellite fleets), although the growing number of actors is forcing increased international competitiveness.

A breakdown of the total shows that while ground equipment grew over 2002-2005, the launch and satellite manufacturing areas shrunk until 2005 (Figure 3.1b). This trend is reflected in the rising percentage of the total revenues coming from the ground segment and proportional declines in both the launch and satellite manufacturing markets (Figure 3.1c).

The US share in revenues of world satellite manufacturers decreased (Figure 3.1e). An examination of European space-related manufacturing units shows a similar picture with sales relatively down since 2000, although picking up in 2006 (Figure 3.1f).Worldwide launch revenues in 2006 had not returned to the high levels of 2000, when the US and other players were vying for launch activity (Figure 3.1d).

Definition

The activities presented here focus exclusively on three segments of satellite industry manufacturing. First is the launch industry segment, which comprises launch services (by private companies for both government and private payloads), vehicle manufacturers and component and subsystem manufacturers. Given the difficulty of separating launch manufacturing from launch services data, both types were included together under the umbrella of launch industry revenues. Second is the satellite manufacturing segment which includes manufacturers of satellites and associated components and subsystems. Third is the ground equipment segment, which covers the manufacturing of mobile terminals, gateways, control stations, VSATs and DBS dishes, and handheld phones and other equipment. Eurospace data focus on the manufacturing of space hardware and software, ground stations, launch equipment and associated parts throughout Europe.

Methodology

The data come primarily from two US Satellite Industry Association (SIA) reports, which are based on surveys that target large companies operating in the three segments, focusing on their employment and revenue situations. The data are complemented with publicly available information to provide a more comprehensive overview of the segment and industry. The launch industry data include information from private companies on both their commercial and non-commercial payloads, but exclude government launches (*e.g.* the Space Shuttle or the International Space Station). The launch industry and satellite manufacturing data are counted on the year of the launch, and all data are presented in current US dollars (they are not adjusted for inflation).

Supplementary data include industry reports from Eurospace, which examine turnover of the European space industry. Eurospace draws on surveys complemented with additional data. The Eurospace data covers the development and production of spacecraft, launchers and related ground equipment.

Data comparability

As mentioned previously, it is a major challenge to obtain comparable international data on space activities, mainly because of statistical classification issues and the limits on the current definition of the industry. Although extensive, the data from private sources raise issues of double counting, especially with regard to revenue statistics (the input of one company may include the output of another one in its total). Some satellite manufacturing revenues may be also slightly misleading, since they reflect revenues when satellites are actually launched (*e.g.* SIA report), with 2005 figures reflecting mostly 2002 orders, a bad year for the industry. In fact in 2005, more than 20 satellites were ordered, so manufacturing revenues are sure to pick up in data from following years.

Data sources

- Satellite Industry Association (2006), *State of the Satellite Industry Report*, Futron Corporation, June.
- Satellite Industry Association/Futron (2007), *State of the Satellite Industry Report*, Futron Corporation, June.
- ASD-Eurospace (2007), *Facts and Charts: The European Space Industry in 2006*, June.

 THE SPACE ECONOMY AT A GLANCE – ISBN 978-92-64-03109-8 – © OECD 2007

3.1. REVENUES FROM THE SPACE INDUSTRY

Figure 3.1a. **World satellite industry manufacturing revenues, 2000-2006**

Billions of US dollars

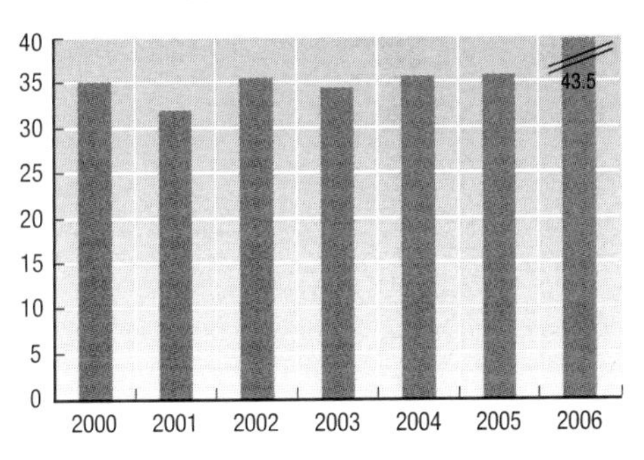

Source: SIA (2007), *State of the Satellite Industry Report*, June.

Figure 3.1b. **World satellite industry manufacturing revenue by sector, 2000-2006**

Billions of US dollars

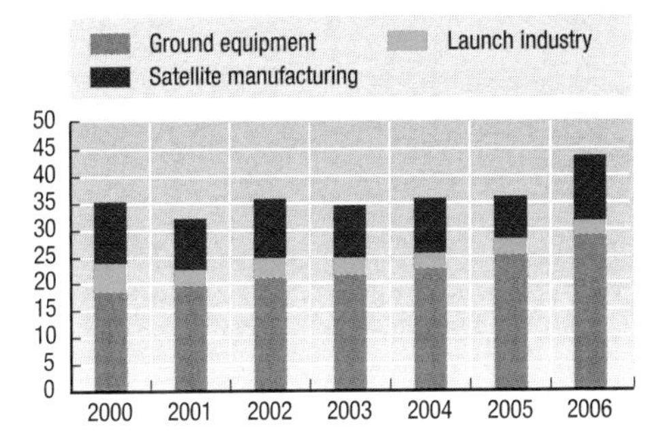

Source: SIA (2007), *State of the Satellite Industry Report*, June.

Figure 3.1c. **World satellite industry manufacturing revenues by sector, 2000-2006**

Percentage of worldwide manufacturing industry revenue

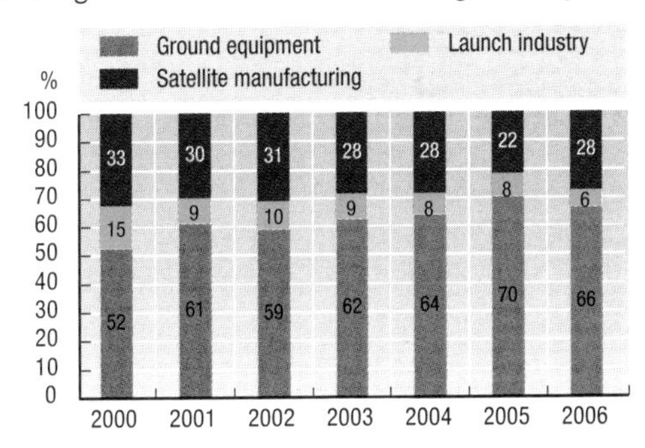

Source: SIA (2007), *State of the Satellite Industry Report*, June.

Figure 3.1d. **Worldwide launch industry[1] revenues, 2000-2006**

Billions of US dollars

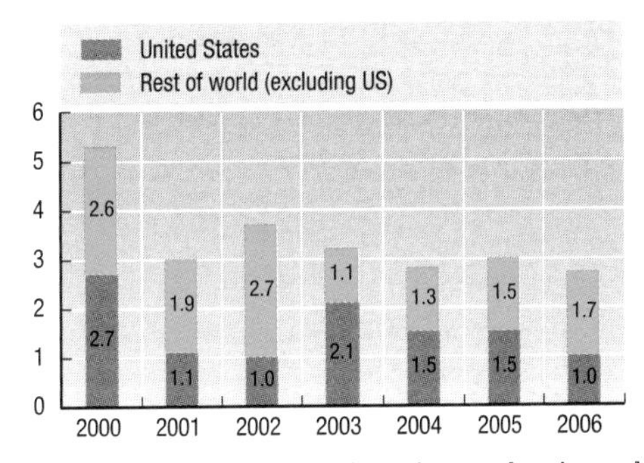

1. Includes both launch manufacturing and private launch services.Revenues based on year of launch and not when contract awarded.

Source: SIA (2007), *State of the Satellite Industry Report*, June.

Figure 3.1e. **Worldwide manufacturing of satellite revenues, 2000-2006**

Billions of US dollars

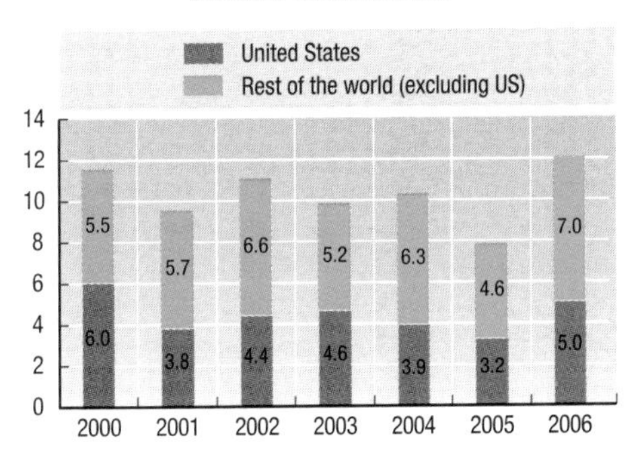

Source: SIA (2007), *State of the Satellite Industry Report*, June.

Figure 3.1f. **Turnover by European space manufacturers, 1992-2006**

Millions of euros

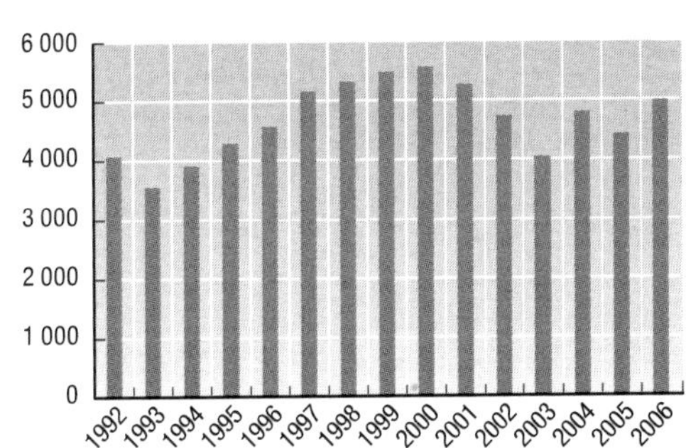

Source: Eurospace (2007), *Facts and Figures*, June 2007.

3.2. SPACE-RELATED SERVICES

While the manufacturing and launch segments (the "upstream" segment) of the space sector have faced some challenges over the past five years (including a levelling-off in demand for launch services), space-related services markets (the "downstream" segment) continue to grow strongly.

Highlights

Space-related services revenues are not easy to gauge nationally and internationally, but worldwide estimates range from some USD 52.2 billion to 77.2 billion in revenues in 2005 (Figures 3.2aand 3.2c).

According to the US Satellite Industry Association (SIA), revenues from the world satellite services industry (mainly telecommunications and Earth observation services) were 83% higher in 2005 than five years earlier, and still growing in 2006. Telecommunications services, in particular direct broadcast satellite (DBS) services (*e.g.* satellite television), represent the bulk of commercial revenues with USD 48.5 billion in 2006 (Figure 3.2b). Further growth is expected due to expected satellite operators' consolidations and strong demand worldwide. Other space-related services, in Earth observation and navigation, are not generating as much revenue (Figure 3.2c), although governments, particularly defence departments, increasingly use satellite capacities, as demonstrated by their use of commercial satellite bandwidth (Figure 3.2d).

Several studies point to an increase in satellite services revenues in different markets over the next decade. For example, as a new generation of systems comes online, growth in the mobile satellite services market is expected to be more robust (Figure 3.2e). Concerning Earth observation, satellite imagery should benefit from increasing worldwide demand for geospatial products (*e.g.* weather forecasting) (Figure 3.2f).

Finally, revenue estimates for space-related services may be largely underestimated, as shown by the findings of the United Kingdom's recent space industry mapping exercise. It showed that companies making commercial use of space assets (capacities or products) were often neglected in existing industry surveys (Box 3.2).

Space-related services revenues are not easy to gauge nationally and internationally, but worldwide estimates range from some USD 52.2 billion to 77.2 billion in revenues in 2005 (Figures 3.3aand 3.2c).

According to the US Satellite Industry Association (SIA), revenues from the world satellite services industry (mainly telecommunications and Earth observation services) were 83% higher in 2005 than five years earlier, and still growing in 2006. Telecommunications services, in particular direct broadcast satellite (DBS) services (*e.g.* satellite television), represent the bulk of commercial revenues with USD 48.5 billion in 2006 (Figure 3.2b). Further growth is expected due to expected satellite operators' consolidations and strong demand worldwide. Other space-related services, in Earth observation and navigation, are not generating as much revenue (Figure 3.2c), although governments, particularly defence departments, increasingly use satellite capacities, as demonstrated by their use of commercial satellite bandwidth (Figure 3.2d).

Several studies point to an increase in satellite services revenues in different markets over the next decade. For example, as a new generation of systems comes online, growth in the mobile satellite services market is expected to be more robust (Figure 3.2e). Concerning Earth observation, satellite imagery should benefit from increasing worldwide demand for geospatial products (*e.g.* weather forecasting) (Figure 3.2f).

Finally, revenue estimates for space-related services may be largely underestimated, as shown by the findings of the United Kingdom's recent space industry mapping exercise. It showed that companies making commercial use of space assets (capacities or products) were often neglected in existing industry surveys (Box 3.2).

Definition

Space-related services use a specific satellite capacity, such as bandwidth or imagery, as inputs to provide a more global service to business, government or retail consumers. Those services are as diverse as space applications themselves. The services are traditionally divided into three large application domains: telecommunications, Earth observation (also called remote sensing) and navigation. Value chains often involve public agencies as investors and final users. As such, public authorities remain significant customers even in well-established commercial markets, such as telecommunications.

Methodology

As there are no official sources providing international estimates for space-related services, private sources have been used to provide at least some orders of magnitude. The activities presented here show the diversity of space-related services, and also the differing methodologies used to assess these markets.

Different definitions of specific products and services co-exist, and some reports from private sources tend to aggregate categories of services (*e.g.* SIA includes some remote sensing services in the category of fixed satellite services). In addition, companies may be nvolved in and gain revenues from various segments of service's value chain. Hence, there are widely ranging esults from US and European reports.

Due to their relative novelty and the lack of existing ata, some space-related services are not covered here. his includes "space tourism" (*i.e.* paying for space dventure rides), which is just starting to develop.

)ata comparability

Commercial satellite services markets are not only ery diverse in nature, but also fragmented nternationally into specific regional markets. Thus, an verarching global view of the space-related services ector is currently difficult to establish.

As in the case of space manufacturing, nternational data from private sources raise issues of ouble counting; and revisions to annual reports have ntroduced notable changes to estimated trends. For xample, the 2006 SIA industry indicators report ignificantly revised its 2004 numbers, with DBS elevision revenues adjusted downward by JSD 13 billion (almost 40% of the total).

However, as demonstrated by the recent space ndustry mapping exercise in the United Kingdom, some ervices markets are still very much underestimated Box 3.2). More work is needed to capture better space-elated services and companies, which often have no irect links to the traditional space sector, but which nevertheless use space components.

)ata sources

- BCC Research (2007), Remote Sensing Technologies and Global Markets, March.
- Bierett, R. (2007), Presentation for Telecom Info Days 2007, European Space Agency, ESTEC, April (using data from Euroconsult, 2006).
- BNSC (2006), BNSC Space Sector Mapping Study, April.
- National Space Society (2006), The Space Report, autumn.
- Northern Sky Research (2006), Government and Military Market for Commercial Satellite Services, March.
- Northern Sky Research (2006), Mobile Satellite Services, second edition.
- Satellite Industry Association (2007), State of the Satellite Industry, Futron Corporation, June

Figure 3.2a. **World satellite industry revenues for services and other[1], 2000-2006**

Billions of US dollars

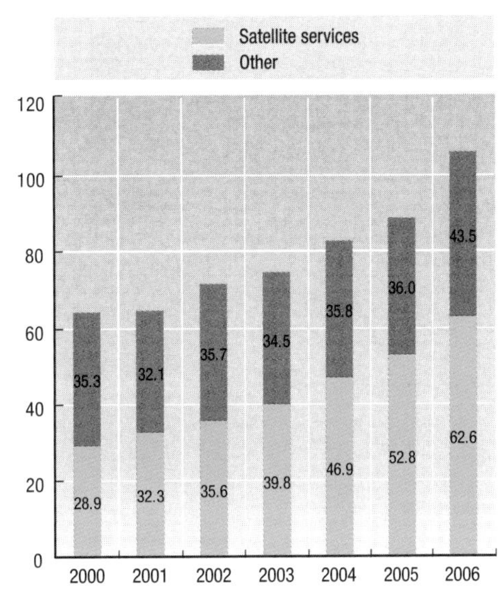

1. "Other" is ground equipment, launch industry and satellite manufacturing.

Source: SIA (2007), *State of the Satellite Industry Report*, June.

Figure 3.2b. **World satellite services revenue, 2000-2006**

Millions of US dollars

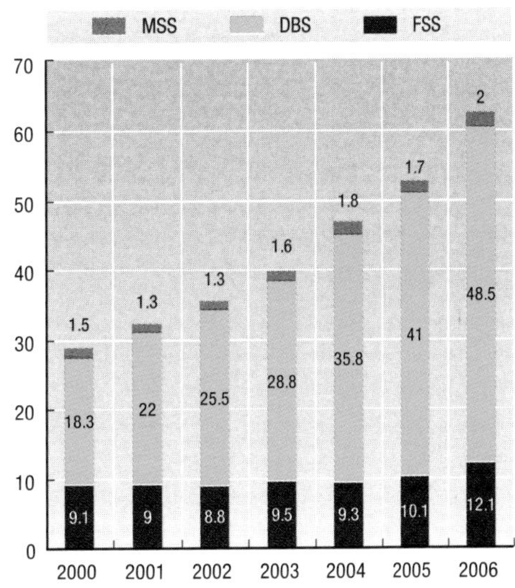

Notes: MSS (Mobile Satellite Services): Mobile telephone and mobile data.

DBS (Direct Broadcast Satellite): Direct to home television (DTH), Digital Audio Radio Service (DARS), and broadband.

FSS (Fixed Satellite Services): Very Small Aperture Terminal (VSAT) services, remote sensing, and transponders agreements.

Source: SIA (2007), *State of the Satellite Industry Report*, June.

3.2. SPACE-RELATED SERVICES

Figure 3.2c. **The three value chains in commercial satellite applications in 2005**

Revenues in billions of US dollars

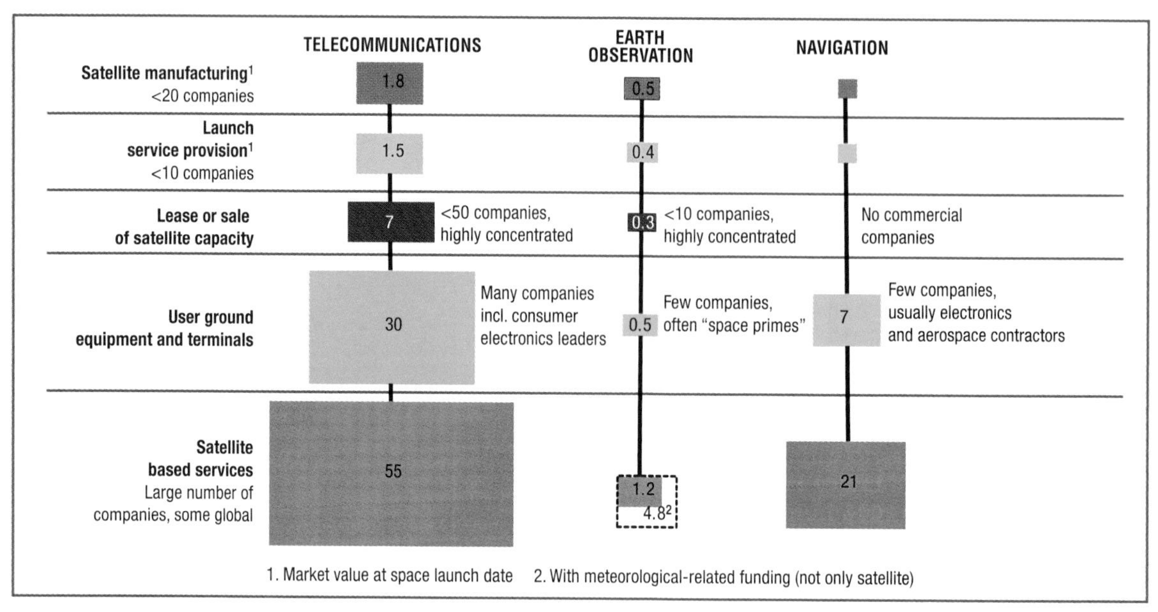

1. Market value at space launch date.
2. With meteorological-related funding (not only satellite).

Source: R. Bierett (2007), *Presentation for Telecom Info Days 2007*, European Space Agency, ESTEC, April (data from Euroconsult, 2006).

Box 3.2. **Lessons learned in estimating space-related services revenues: The 2006 UK industry mapping study**

In 2005-2006, the British National Space Centre carried out an industry mapping study to help inform the UK Civil Space Strategy 2007-2010. A thorough study of the supply chains and networks of value added in the UK space sector helped identify a number of players that were not included in previous space industry surveys (i.e. BNSC's *Annual State and Health of the Space Sector* reports). It was found that there are many downstream application areas and markets in which space technologies are significant enablers and which represent large amounts of turnover. According to the research conducted, companies which sell satellite broadcast receivers as well as companies which resell satellite navigation transponders and satellite phones could / should be included if they make a partial use of space data in their business. In that context, the value-added figures from the *Annual State and Health of the Space Sector* report of GBP 2.2 billion are judged too small by at least 25%.

Figure 3.2d. **World government and military commercial satellite market total, 2003-2012**[1]

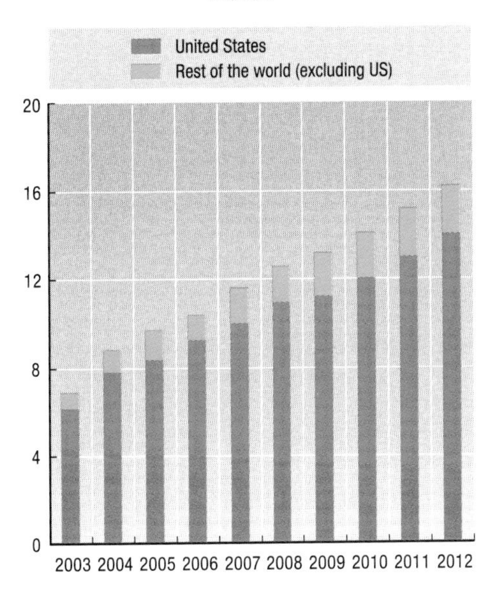

Bandwidth procured in gigabits per second

1. Estimated 2007-2012.

Source: Northern Sky Research (2006), *Government and Military Market for Commercial Satellite Services*, March.

Figure 3.2e. **World mobile satellite services market: Wholesale and retail revenues, 2003-2012**[1]

Billions of US dollars

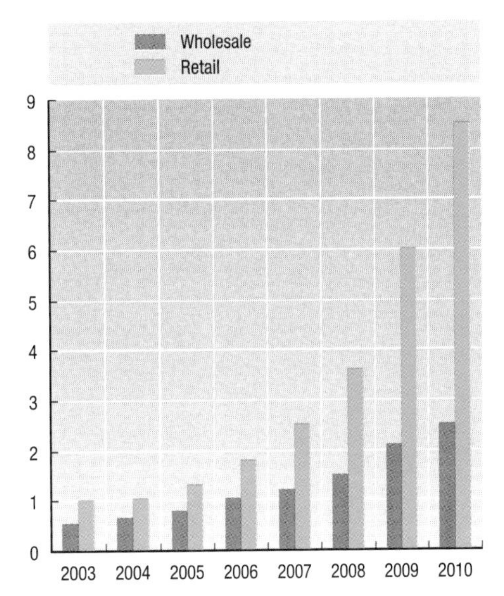

1. Estimated 2007-2012.

Source: Northern Sky Research (2006), *Mobile Satellite Services*, second edition.

Figure 3.2f. **Estimated global expenditures for remote sensing products by application, 2006-2012**[1]

Millions of US dollars[2]

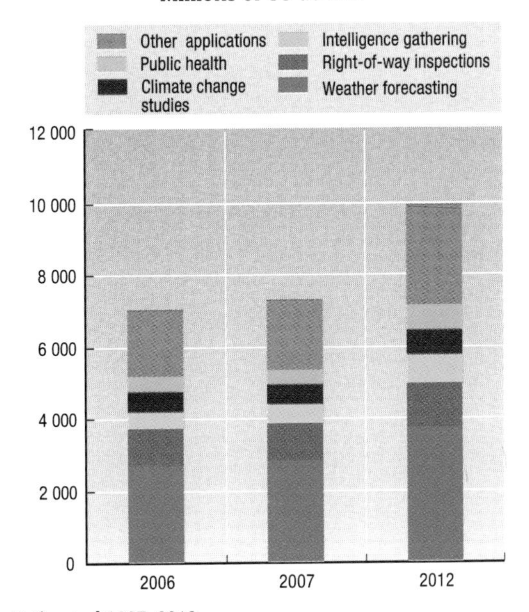

1. Estimated 2007, 2012.
2. Includes more than just satellite imagery, i.e. aerial.

Source: BCC Research (2007), *Remote Sensing Technologies and Global Markets*, March.

3.3. INTERNATIONAL TRADE IN SPACE PRODUCTS

Although not many space products and services are fully commercial (*i.e.* most are strategic in nature and not freely traded), this section provides a partial overview of existing trade data by examining the exports of two commodity groups with significant space components from the International Trade in Commodity Statistics (ITCS) database, defined in detail below. Exports are those of OECD countries. Note that the United Nations refers to the database as the COMTRADE (Commodity Trade Statistics) database.

Highlights

Data for 2004 from OECD member countries reveal that exports of space products are dominated by a few major countries, with the G7 accounting for 91%, and the US, France and Germany alone accounting for 71% (Figure 3.3a).

Total space exports in 2004 fell by 13% compared with 2003, to USD 3.74 billion (Figure 3.3b). While exports of "*Spacecraft, including satellites, and suborbital and spacecraft launch vehicles*" (ITCS category HS880260) rose by USD 570 million, exports of the much larger category "*Parts of balloons, dirigibles, and spacecraft not elsewhere specified*" (HS880390), dropped by USD 1.135 billion leading to an overall decline from 2003 of USD 560 million.

Statistics from 1996 to 2004 also show that recent exports by OECD countries have fallen substantially from their high of 1998. It is important to note that a majority of exports since 1998 have been in the commodity code that includes non-space items ("*Parts of... spacecraft*") in addition to "*Spacecraft, including satellites...*". This trend parallels the cyclical downturn of the aerospace sector around 2001, mentioned previously.

An examination of G7 exports for 2004 reveals that they are focused on a few key markets (Table 3.3). In fact, 97% of the USD 3.395 billion space exports went to three of the 10 continents/regions and intra-G7 exports accounted for USD 1.98 billion (58%) of the total. Among non-OECD markets, Asia appears to be the most important with over 75% of the total.

Definitions

Trying to determine what exactly constitutes trade in space-related commodities can be complicated. Nevertheless, the two commodity codes employed clearly indicate "space-related" elements: (1) HS880260 ("*Spacecraft, including satellites, and suborbital and spacecraft launch vehicles*"); and (2) HS880390 ("*Parts of balloons, dirigibles, and spacecraft not elsewhere specified*"). The estimates from those commodity codes include therefore more items than just space products.

Methodology

Statistics on the quantity and markets for exports of OECD economies come from the ITCS database jointly managed by the OECD and the United Nations. It includes details on imports and exports for all UN member states. The OECD is responsible for the collection of statistics related to its member countries and the UN for all others. Exports from these two commodity classifications are those of OECD countries to all countries of the world (both OECD and non-OECD member countries).

Comparability

The only case where export data on these commodities are not available is for the United Kingdom which lacked the data for 1999, 2000, 2003 and 2004. As substitute measures of UK exports for these missing years, the values of imports of these two commodities by the rest of the world from the UK were used. All statistics are presented in current US dollars, by converting domestic currencies using annual trade-weighted aggregates of monthly exchange rates.

Data sources

- OECD / UN (2007), International Trade in Commodity Statistics (ITCS) database, April.

THE SPACE ECONOMY AT A GLANCE – ISBN 978-92-64-03109-8 – © OECD 2007

Table 3.3. **G7 total exports of space products by country of destination,[1] 2004**

Millions of current US dollars of exports

2004 G7 totals	Millions of US dollars	Percent
TOTAL	3 394.56	100.0%
By continent/region of destination:		
Europe	1 662.84	49.0%
Asia	1 236.36	36.4%
North America	395.36	11.6%
South America	45.78	1.3%
Middle East	21.47	0.6%
Africa	18.39	0.5%
Oceania	11.50	0.3%
Central America	2.94	0.1%
Unspecified	0.032	0.0%
Antarctica	0.000	0.0%
Of which:		
OECD countries	2 432.24	71.7%
Non-OECD countries	962.29	28.3%
Unspecified	0.00	0.0%
OECD countries	2 432.24	100.0%
Of which:		
G7 Countries	1 979.47	81.4%
Non-OECD countries	962.29	100.0%
of which:		
Asia (excluding Middle East)	724.26	75.3%
Europe	149.34	15.5%
Americas	48.73	5.1%
Middle East	21.47	2.2%
Africa	18.39	1.9%
Oceania	0.10	0.0%

1. Space products are: HS880260 (Spacecraft, etc.) and HS880390 (Parts of balloons, spacecraft, etc.).

Source: OECD/UN (2007), *International Trade in Commodity Statistics (ITCS) database, April.*

Figure 3.3a. **Amount and share of OECD space products exports, 2004**

Exports in millions of current US dollars and as a percentage of OECD total

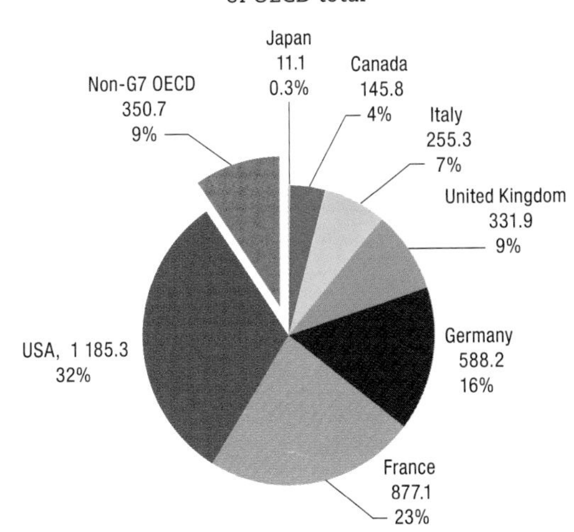

Source: OECD/UN (2007), International Trade in Commodity Statistics (ITCS) database, April.

StatLink http://dx.doi.org/10.1787/105381478683

Figure 3.3b. **OECD Exports of Space Products 1996-2004**

Exports in billions of current US dollars

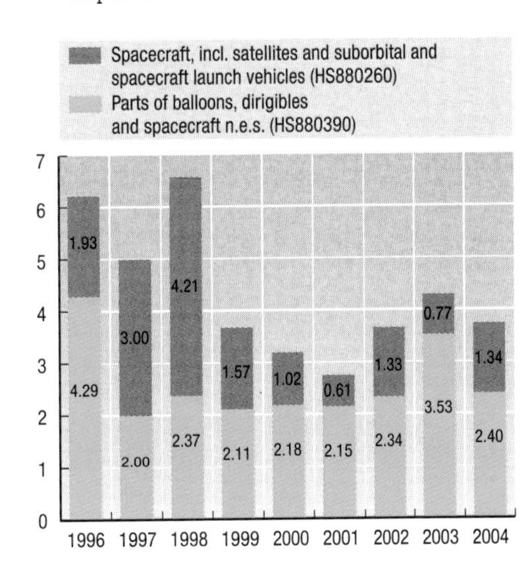

Source: OECD/UN (2007), International Trade in Commodity Statistics (ITCS) database, April.

StatLink http://dx.doi.org/10.1787/105405043773

3.4. SPACE PATENTS

Over the years, OECD work has shown the reliability of patent data as an indicator of the technological innovation and economic health of a given sector. This is also true for the space sector and its derived applications. Surveys show that a large proportion of firms' inventions are patented and that a large proportion of patents become innovations with an economic use. Patents reveal inventions and innovations in small firms and in the engineering departments of large firms, which R&D indicators alone do not properly measure.

Highlights

Ever since the appearance of the first satellites and other systems at the dawn of the space age in the late 1950s, the space sector has been in the forefront of high tech innovation. More recently, the convergence of new information technologies and computer power has benefited both space systems and innovative "down-to-Earth" applications (*e.g.* communications, navigation, imagery). The number of space-related patents tripled between 1990 and 2000, both in Europe and the United States, but declined from 2001 on, due to a large degree to time-lag effects described below (Figures 3.4a and 3.4b).

Between 1980 and 2004, the OECD countries were responsible for 97% of all space-related applications to the European Patent Office (EPO) and nearly all the grants at the United States Patent and Trademark Office (USPTO) (Figures 3.4c and 3.4d). The US was the largest applicant with 47% of space patents at the EPO and 75% at the USPTO. France, Germany and Japan also accounted for a major portion of space-related patents at both offices. While the US tended to focus on "*Cosmonautics; vehicles or equipment thereof*" (category B64-related in the international classification) other countries (especially Japan) tended to focus on other more specialised patents (Figures 3.4e and 3.4f).

Definition

The space-related patents referred to in the figures primarily include all systems and applications included in the international statistical classification B64G: "*Cosmonautics; vehicles or equipment thereof*". This classification covers a large array of space-related systems and applications (including satellites; launchers; components; radio or other wave systems for navigation or tracking; simulators). In addition, a few other patent classifications were included, provided that the patent description contained certain key words.[1]

Methodology

For this analysis, space-related patents are defined using a mixture of International Patent Classification (IPC) codes and keywords. The principle IPC class used is "B64G" ("*Cosmonautics; vehicles or equipment thereof*") which covers technology related to developing and maintaining space-based systems, space exploration and peripheral equipment related to cosmonautics. The simplest type of patent indicator is derived by counting the number of patents that satisfy certain criteria. The criterion here was either that the statistical classification be "B64G" or another relevant patent classification containing certain keywords. In order to capture patents relating to applications relying on space-based technology, patent applications with the following IPC classes were chosen if the title of the patent application contained one or more of the following phrases: "GPS", "global position", "satellite", "remote sensing", "earth observation" and "geographic information system": *a)* G01S – Radio direc tion-finding; radio navigation; determining distance or velocity by use of radio waves; locating or presence-detecting by use of the reflection or re-radiation of radio waves. *b)* H01Q – Aerials. *c)* Radio transmission systems: H04B7/185/19 and /195 – Space-based or airborne stations, earth-synchronous stations and non-synchronous stations.

The data came from the OECD Patent Database, which provides links to all major patent databases such as those of the EPO and USPTO.

A key methodological issue is related to the visible downturn of patent applications since 2001. This is mainly due to delays and technical difficulties in updating patent databases and also the time-lag at the USPTO between the application of a patent and its granting. Thus, the downturn should not be misconstrued as a recession in terms of space-related patenting activities. Work is ongoing to see if space patents can be examined in greater detail to determine, for example, the linkages between patents and citations, licenses and other gauges to help quantify the relationship between patents and product development.

Data comparability

Patents presented here do not cover all space-related innovations, as many are protected by other types of intellectual property regimes, or by secrecy. The patenting activity of individual countries may also differ widely, depending on the patent institution considered. National data on countries' patenting activity can be broken down by region in order to investigate the geographical distribution of technological activities. The main methodological problem is how to assign individual patents to regions in a way that reflects the presence of inventive activity, as patents are usually assigned according to the address of the inventor or the firm that owns the patent.

Data sources

- OECD (2006), *OECD Patent Database*, September.

Figure 3.4a. **Breakdown of space-related patents at EPO, 1980-2003**[1]

Number of patents granted or pending by country of applicant

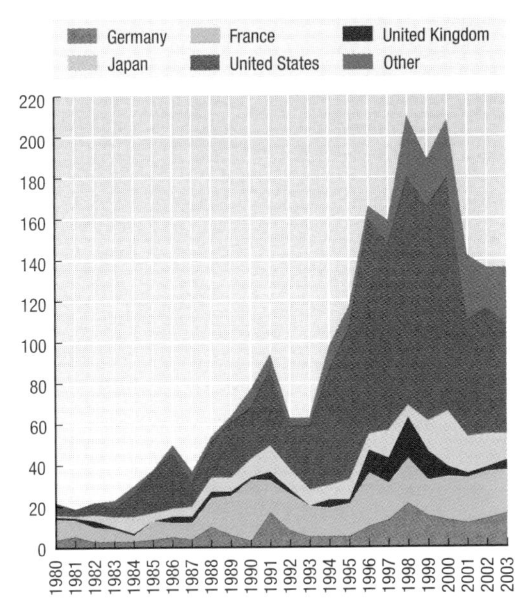

1. Please note the impact of time lag on last few years of data.

Source: OECD (2006), *OECD Patent Database*, September.

StatLink http://dx.doi.org/10.1787/105406846328

Figure 3.4b. **Breakdown of space-related patents granted at USPTO, 1980-2002**[1]

Number of patents granted per year to applicants from all countries

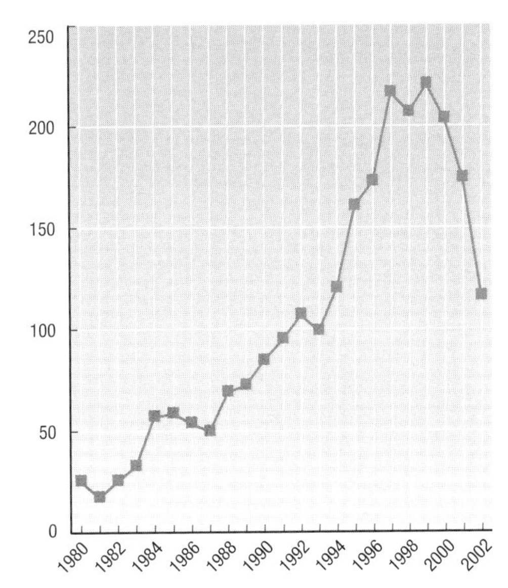

1. Please note the impact of time lag on last few years of data.

Source: OECD (2006), *OECD Patent Database*, September.

StatLink http://dx.doi.org/10.1787/105414700135

Figure 3.4c. **Breakdown of space-related patenting at EPO, 1980-2004**

Percent of all patents based on country of inventor

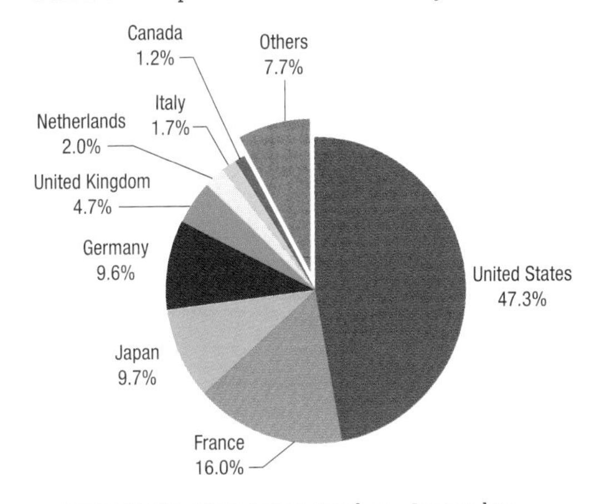

Source: OECD (2006), *OECD Patent Database*, September.

StatLink http://dx.doi.org/10.1787/105503285727

Figure 3.4d. **Breakdown of space-related patenting at USPTO 1980-2004**

Percentage of all patents based on country of inventor

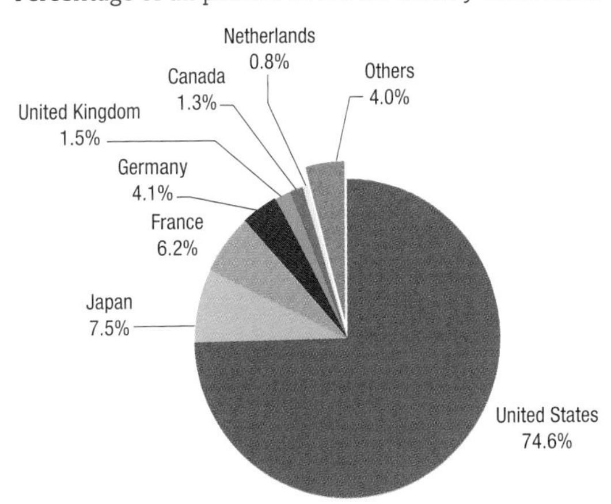

Source: OECD (2006), *OECD Patent Database*, September.

StatLink http://dx.doi.org/10.1787/105556480168

3.4. SPACE PATENTS

Figure 3.4e. **Breakdown of space-related patents by type and country at EPO, 1980-2004**

Number of patents granted or pending based on country of inventor

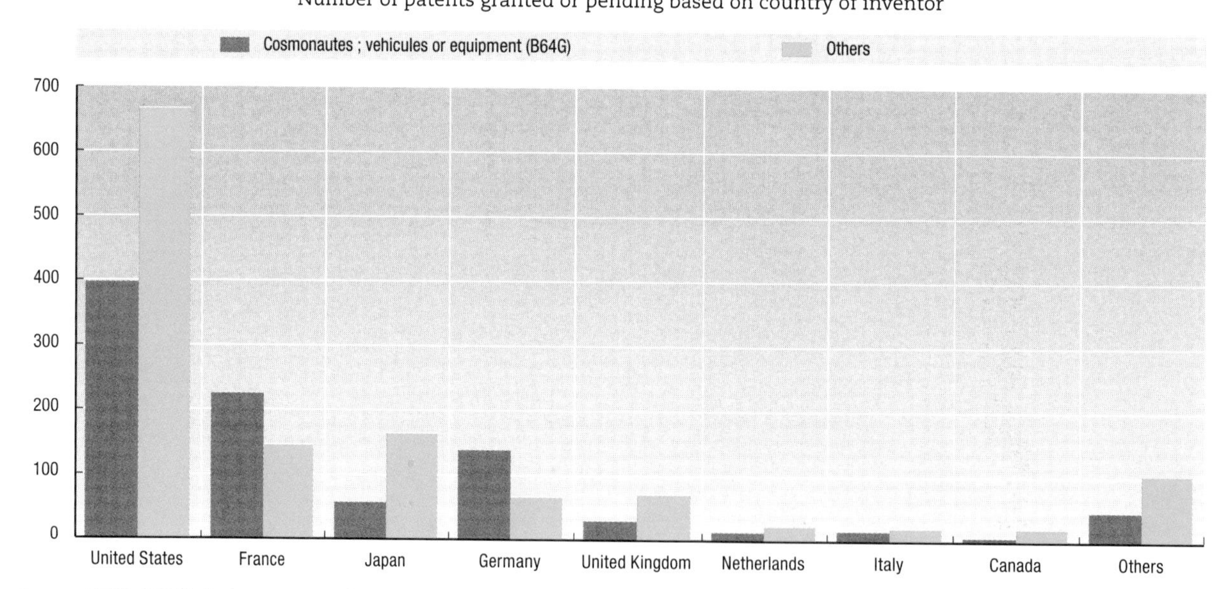

Source: OECD (2006), *OECD Patent Database*, September.

StatLink http://dx.doi.org/10.1787/105600723725

Figure 3.4f. **Breakdown of Space-related Patents by Type and Country at USPTO, 1980-2004**

Number of Patents Granted Based on Country of Inventor

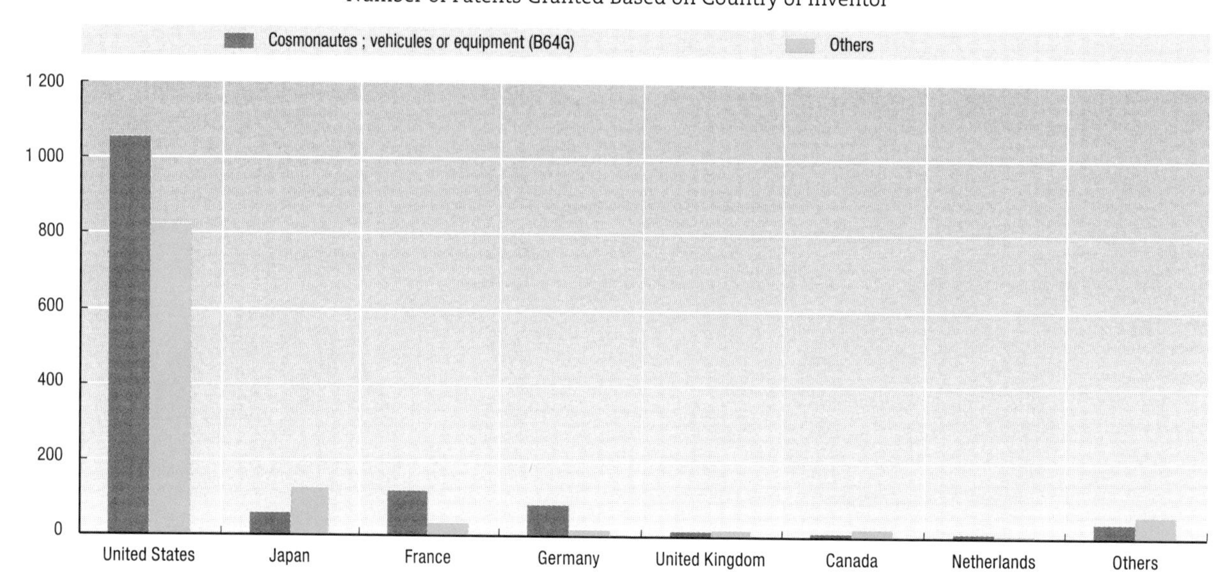

Source: OECD (2006), *OECD Patent Database*, September.

StatLink http://dx.doi.org/10.1787/105657466582

3.5. SPACE LAUNCH ACTIVITY

A dozen countries currently have an autonomous capability to launch satellites into orbit. The international space launch industry plays a pivotal role in enabling commercial and non-commercial actors to engage in civilian and military space activities.

Highlights

The number of launches has fallen off since the late 1990s (Figure 3.5a). Commercial launches have decreased largely due to the financial crisis faced by telecom operators in 2001. Not surprisingly, the same pattern is displayed in payload activity (Figure 3.5b). An examination of all launches by country from 2000 to 2006 reveals that, while all major launch providers (US, Russia and Europe) had fewer launches in 2006 than in 2000, some of the decline in the US and Europe was offset by gains by Russia and China (Figure 3.5c).

An examination of commercial launch events only over 1996-2000 and 2001-2006 reveals that gains by both Russia and the multinational firm Sea Launch happened at the expense of China and the US (Figures 3.5d and 3.5e). Revenues from commercial launches have tended to decline with declining launch activity (Figure 3.5f). The cyclical nature of satellite activities (*i.e.* the need to renew satellite fleets) and the growing number of countries with space programmes should contribute to more space launches over the next decade. International competition in commercial markets is likely to increase.

Definition

Space launch events can be broken down into two main types: commercial and non-commercial. A commercial launch event is one where the primary payload's launch is open to competition from any capable launch service provider. Hence, a commercial launch may be performed by either a government or private launch service provider. Conversely, a non-commercial launch event is any launch activity where the orbital transport service of the primary payload is not subject to competition.

Launch events from the Sea Launch venture refer to "multinational" launches that are done in international waters involving the partnership of organisations from four different countries (Norway, Russia, the Ukraine and the United States).

The payload may include one or more satellites, and may also be commercial or non-commercial. Commercial payloads refer to those where either: (1) the payload operator is a private firm; or (2) the payload is government-funded but it provides partial or total services through a semi- or totally private company. Non-commercial payloads can be of civil or military/government nature or not-for-profit (e.g. scientific exploration probes).

Methodology

The data included are mainly provided by the Federal Aviation Administration's Office of Commercial Space Transportation (FAA/AST). They include worldwide orbital and sub-orbital launch events that are conducted during a given calendar year (regardless of when the contract was signed). The data include all launch events and payloads, whether or not the launch or mission is considered to have succeeded.

Data comparability

The FAA data are subject to revision because of subsequent reclassification of commercial/non-commercial launches. Data on total launches were compared to data available from NASA's *Aeronautics and Space Report of the President, Fiscal Year 2004 Activities* which showed NASA having on average just two more launches per year (from 1997 to 2003) than FAA data. Other industry reports from different sources might vary in their definitions of commercial launches, but generally provide the same types of data on launch events.

Data sources

- Federal Aviation Administration's Office of Commercial Space Transportation (FAA/AST) (2007), *Commercial Space Transportation 2006: Year in Review*, January.

3.5. SPACE LAUNCH ACTIVITY

Figure 3.5a. **Total commercial and non-commercial launch events 1998-2006**

Number of launch events

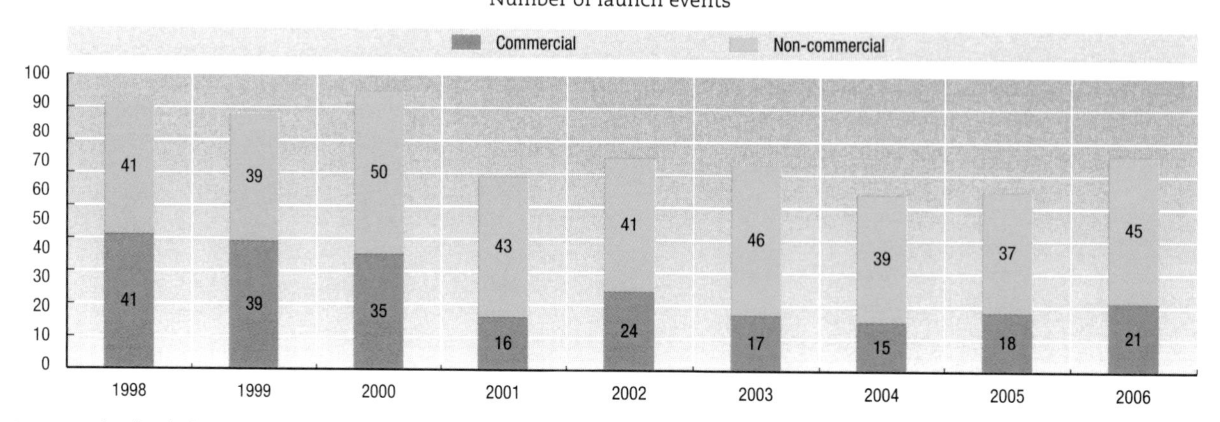

Source: Federal Aviation Administration's Office of Commercial Space Transportation (FAA/AST) (2007), *Commercial Space Transportation 2006: Year in Review*, January.

Figure 3.5b. **Total worldwide commercial and non-commercial payloads, 1998-2006**

Number of payloads

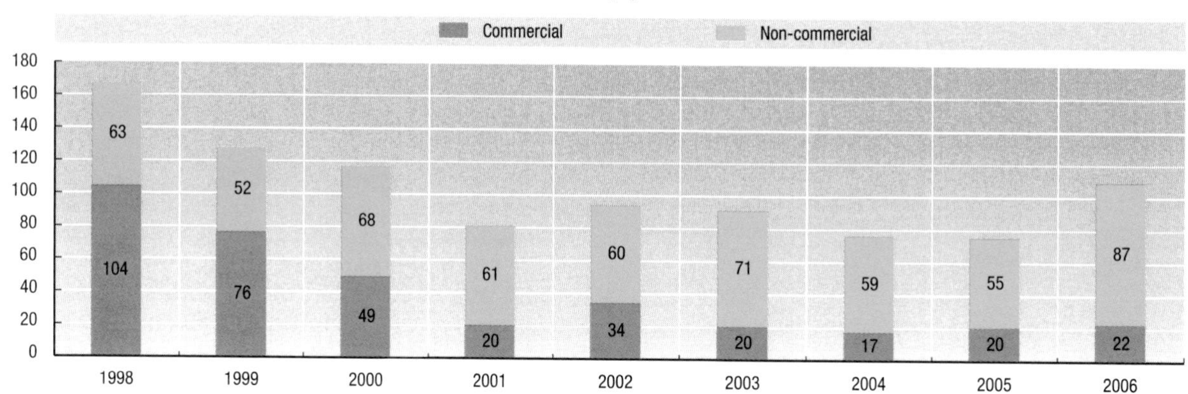

Source: Federal Aviation Administration's Office of Commercial Space Transportation (FAA/AST) (2007), *Commercial Space Transportation 2006: Year in Review*, January.

Figure 3.5c. **Total (commercial and non-commercial) launch events by country, 2000-2006**

Number of launch events

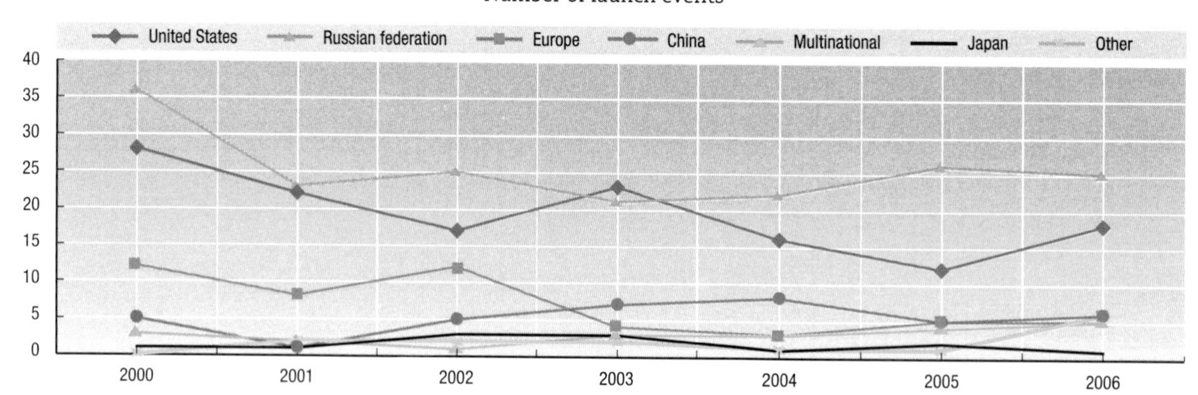

Source: Federal Aviation Administration's Office of Commercial Space Transportation (FAA/AST) (2007), *Commercial Space Transportation 2006: Year in Review*, January.

Figure 3.5d. **Breakdown of 177 worldwide commercial launch events, 1996-2000**

As a percentage of all launch events

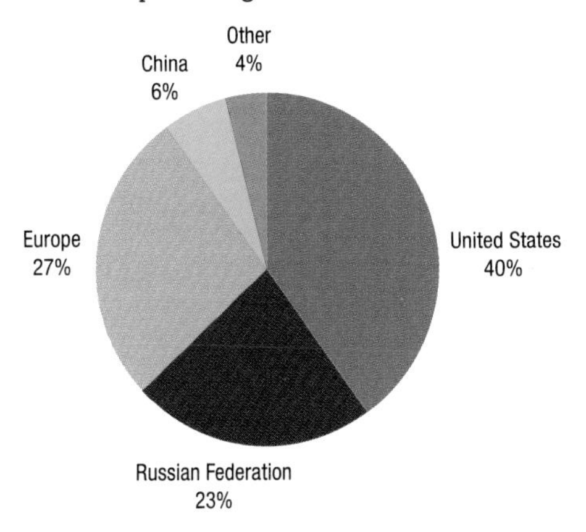

Source: Associate Administrator for Commercial Space Transportation (AST), 2001.

Figure 3.5e. **Breakdown of 111 worldwide commercial launch events, 2001-2006**

As a percentage of all launch events

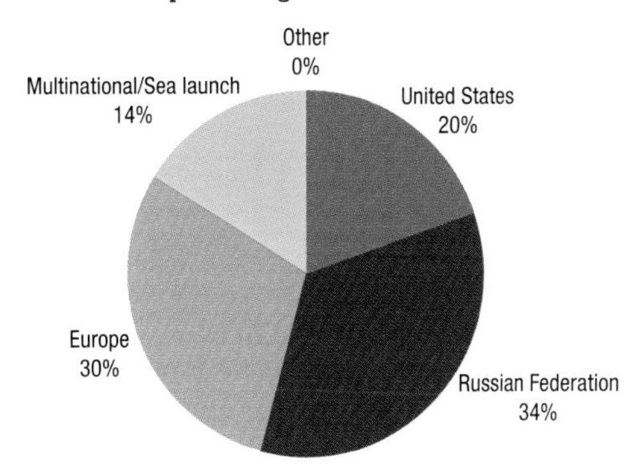

Source: Federal Aviation Administration's Office of Commercial Space Transportation (FAA/AST) (2007), *Commercial Space Transportation 2006: Year in Review*, January.

Figure 3.5f. **Total worldwide commercial launch events and revenue, 1997-2006**

Number of launch events and launch revenues

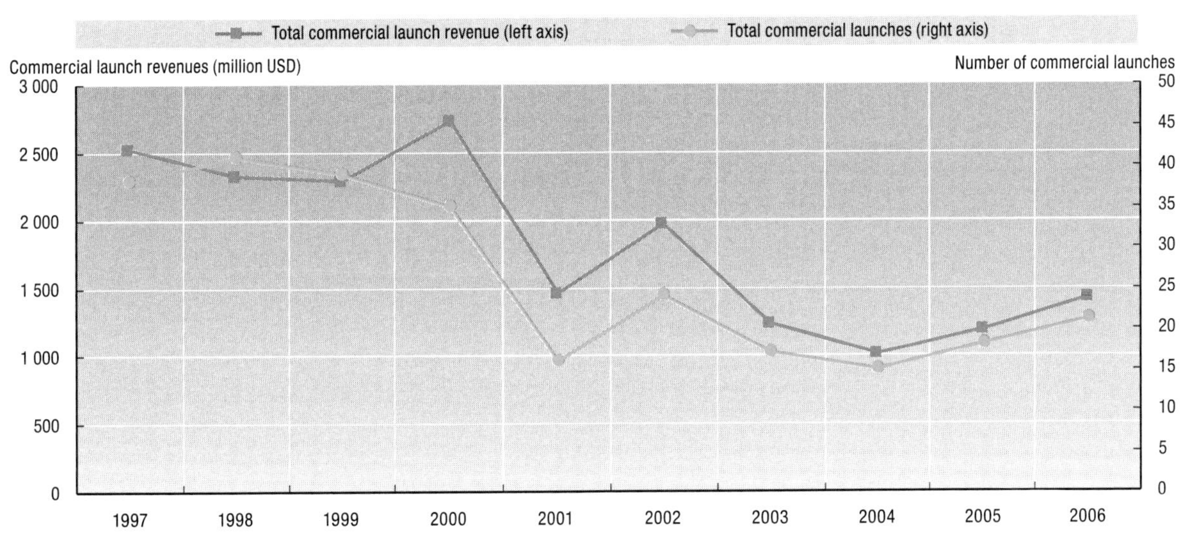

Source: Federal Aviation Administration's Office of Commercial Space Transportation (FAA/AST) (2007), *Commercial Space Transportation 2006: Year in Review*, January.

3.6. SPACE EXPLORATION-RELATED ACTIVITIES

Countries with space programmes are increasingly investing in "down-to-Earth" space applications (*e.g.* telecommunications, Earth observation) for strategic and economic reasons. Nevertheless, space exploration remains a key driver for investments in innovative R&D and sciences, and it constitutes an intensive activity for major space agencies.

Highlights

Space exploration is probably the most visible face of space activities and constitutes an inherent mission of space agencies worldwide. Its achievements generate enthusiasm among the public and wide media interest, as shown by race to the Moon, Mars exploration by robots or the probe landing on Titan. Space sciences and planetary missions have developed markedly over the years. This trend is reflected in the current and planned robotic exploration missions of the solar system, in which the US, Europe and several Asian countries are active players (Table 3.6a).

In addition to robotic exploration, the development of a human presence in space has been a recurring theme since the 1950s for both political and prestige-related reasons. Currently only three countries – Russia, the US and China – have the autonomous capability to launch human beings into space; however, a total of 451 persons from 37 different countries have flown in Earth orbit as of late December 2006. Since the late 1990s, the feasibility of commercial human spaceflight endeavours is also being tested via "space tourism" ventures (Table 3.6b).

Definition

Space exploration is the physical exploration of outer-Earth objects, via robotic probes and human missions. More broadly, it also includes the scientific disciplines (*e.g.* astronomy, solar physics, astrophysics, planetary sciences), technologies and policies applied to space endeavours.

Methodology

Robotic missions presented here include active and planned orbiters (*i.e.* spacecraft whose purpose is to orbit a planet or an asteroid, usually to map the surface), planetary rovers (*i.e.* robots landing and roving on celestial bodies), and other exploration probes (*i.e.* spacecraft sent to fly by several celestial bodies). Planned missions may be cancelled, therefore only missions intended to be launched by 2008 have been included.Several dozen exploration probes have been launched over the years as national or international missions, targeting planets, moons, comets and asteroids in the solar system.

In the case of human spaceflight, several definitions for "astronaut" co-exist. The International Aeronautic Federation (IAF) calls anyone who has flown at an altitude of 100 kilometres an "astronaut". The US Air Force set the limit at fifty miles altitude (80.45 km), while other organisations consider that a person must have reached orbital velocity and remain in orbit (above 200 km) to be considered an "astronaut". The IAF definition has been used here.

Data comparability

The data presented are compiled from various sources. As there is no single information depository for international space exploration missions, the figures provided are estimates.

Data sources

- OECD IFP research (2007), including data from NASA Space Exploration website *http://solarsystem.nasa.gov*, accessed January; ESA Space Science website *www.esa.int*, accessed January; the online *Astronautic Encyclopediahttp://astronautix.com*, accessed January; and communications from French space experts C. Lardier and P. Coué.

THE SPACE ECONOMY AT A GLANCE – ISBN 978-92-64-03109-8 – © OECD 2007

Table 3.6a. Selected active and upcoming robotic exploratory probes, as of December 2006[1]

Name of mission	Date of launch	Agency(ies)	Mission description
Lunar Reconnaissance Orbiter	2008	NASA (USA)	Lunar orbiter
Chang'e 1 ("Moon Goddess")	2007	CAST (China)	Lunar orbiter
Chandrayaan 1 (Hindi for "Moon Craft")	2007	ISRO (India)	Lunar orbiter
Selene	2007	JAXA, ISAS (Japan)	Lunar orbiter
Dawn	2007	NASA (USA)	Rendezvous and orbit asteroids Vesta (2011) and Ceres (2015).
Phoenix	2007	NASA (USA)	Lander to dig soil on northern plains of Mars and look for water-ice evidence (2008).
New Horizons	19 July 2006	NASA (USA)	On its way to Pluto and Kuiper belt (2015), flyby of Jupiter (2007).
Venus Express	9 Nov. 2005	ESA (Europe)	Venus orbiter
Mars Reconnaissance Orbiter	12 Aug. 2005	NASA (USA)	Mars orbiter
Messenger	2 Aug. 2004	NASA (USA)	On its way to Mercury (2011), flyby of Venus (2007).
Rosetta	2 March 2004	ESA (Europe)	On its way to Comet Churyumov-Gerasimenko (2014), flybys of Asteroid 2867 Steins (2008).
Opportunity	7 July 2003	NASA (USA)	Mars rover
Spirit	10 June 2003	NASA (USA)	Mars rover
Hayabusa ("Peregrine Falcon")	9 May 2003	JAXA, ISAS (Japan)	Landed and collected surface samples from the asteroid Itokawa (2005). Return to Earth planned for 2010.
Mars Express	6 Feb. 2003	ESA (Europe)	Mars orbiter
2001 Mars Odyssey	7 April 2001	NASA (USA)	Mars orbiter
Cassini	15 Oct. 1997	NASA, ESA, ASI (USA, Europe, Italy)	Saturn orbiter (the Huygens probe carried onboard landed on Titan in 2005).
Ulysses	6 Oct. 1990	NASA (USA)	Solar orbiter
Voyager 2	20 Aug. 1977	NASA (USA)	Exploration outside the solar system (currently +12 billion kilometres away from the Sun).
Voyager 1	5 Sept. 1977	NASA (USA)	Exploration outside the solar system (currently +15 billion kilometres away from the Sun).

1. In addition to those robotic exploration missions targeted at extraterrestrial bodies, more than a dozen space science satellites are in Earth orbit. Two large international space telescopes (NASA/ESA) are active as of Dec. 2006: the Hubble Space Telescope (launched in 1990) and SOHO, the Solar and Heliospheric Observatory (launched in 1995). Hubble's successor, the James Webb Space Telescope could be launched in 2013. The international CoRoT observatory, led by the French Space Agency (CNES) (launched in 2006), and NASA's Kepler observatory (to be launched in 2008) are designed in particular to search for Earth-like planets outside the solar system.

Source: OECD / IFP research (2007).

Table 3.6b. Selected human spaceflight statistics as of December 2006

Countries with autonomous capability to launch humans into space	3[1]
Number of launches with humans onboard	+240
Persons who have flown into orbit	451
Persons who have flown over the 100 km altitude threshold (including suborbital flights)	454
Number of nationalities who have flown in space	37
Astronauts who walked on the Moon (1969-1972)	12
Operational and inhabited space stations since the 1960s	9[2]
Professional astronauts currently in orbit (the International Space Station is continuously inhabited since 2003)	3
Number of paying orbital spaceflight participants ("space tourism")	4

1. China, Russia, US.
2. 7 Russian, 1 US, 1 international.

Source: OECD / IFP research, 2007.

4. IMPACTS OF SPACE ACTIVITIES

Data in previous chapters on the inputs and outputs of the emerging space economy illustrate how the use of space assets for various applications seems to be increasing and, with it, the impacts on economy and society. This chapter illustrates various types of impacts derived from the development of space activities, using information from diverse sources. Where possible, this information is quantitative, but more often it is qualitative.

The main message is that many space-based services have positive impacts on society, but issues concerning economic data definitions and methodologies have to be resolved to allow the benefits to be identified and quantified more precisely.

This chapter begins with an overview of the categories of impacts derived from space activities. The following sections illustrate the commercial revenue multiplier effect for non-space sectors; the impacts from space-based services on key societal challenges (the environment and natural disasters); and the more specific impacts of institutional space programmes on space firms.

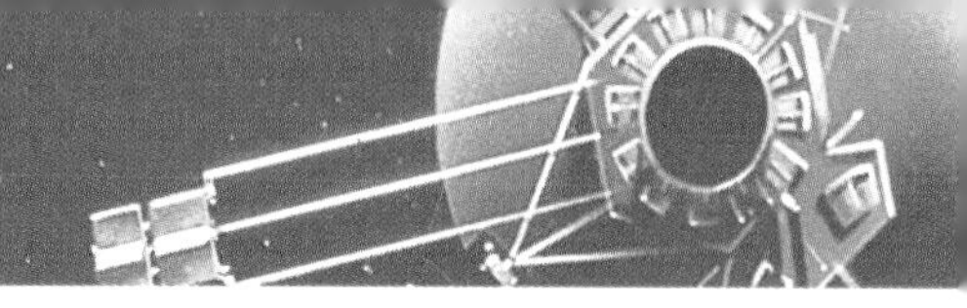

4.1. CATEGORIES OF IMPACTS

The adoption and diffusion of new technologies, such as space technologies, can bring about significant changes, although they may be imperceptible. Important capabilities made possible by space assets include the ability to disseminate information over broad areas, instantaneous telecommunications, and a global vision of the world. However, providing civilian space-based infrastructure, as a useful complement to terrestrial ones, is a relatively recent objective. Space activities were (and still are in some cases) developed primarily for strategic and military purposes, not for economic or societal gains.

This perspective has slowly changed with the growing integration and ubiquitous use of space-based services in various policy making and commercial activities. It is nevertheless acknowledged that science and space exploration, as long-time drivers of innovation in space developments, are the key mission of space agencies as mentioned in OECD (2005) and in Section 3.6 in this book. Today, space assets have diverse impacts on society, although those impacts are generally not well identified and the information available about them tend to be more qualitative than quantitative. The table below summarises several types of impacts. Specific examples are provided in the next sections.

Table 4.1. **Selected types of impact of space investments**

Category of impact	In the space sector	In other sectors
New jobs	Workforce in the space sector	Employment locally, regionally serving the space sector workforce (*e.g.* local shops, industries).Employment in companies, organisations, using space-related products or services to create new products or services (*e.g.* imagery in geospatial equipment, satellite signals in navigation equipment).
New revenues	Revenues from new services	Revenues coming from new services, based on space-based elements (telecommunications, navigation, geospatial services).
Efficiency	Increased competitiveness of some space firms (see Norway example)	Productivity gains achieved by improving space assets users' production and distribution. Cost savings.
Cost avoidance	—	Reduced damage to properties and lives.
Social inclusion	—	Satellite communications infrastructure projects contribute to addressing the problem of social exclusion by improving accessibility.

Box 4.1. **Methodological challenges in impacts analysis**

Space activities, embedded in the larger aerospace domain, can be a noteworthy contributor to an economy. The main issue is to provide reliable measures of the impacts, and this requires sound background data. As mentioned in previous sections, the current supply of space-related statistics contains many gaps. Many of the currently available input metrics are of low comparability across countries or of limited availability. This is part of a general problem, that is, the most widely available indicators of space activities are the least useful for tracking the development of a "space economy" (*e.g.* services), whereas the most valuable indicators are largely missing, except for a few applications. There is a lack of official space-related data for calculating true productivity based on value added per employee from the sales of space goods and services, for example. The effect of space on competitiveness is also hard to determine, since it would require data on the cost of using specific space production processes or the profitability of specific space products.

4.2. COMMERCIAL REVENUE MULTIPLIER EFFECT FOR NON-SPACE

A number of studies suggest that space activities can have significant economic impacts on other sectors, although the revenue multiplier effect often takes place only after years of R&D have led to operational space systems.(Of course a certain amount of investment stays in the space sector (Bach, 2002), due to the development of items that can only be used internally or for programmes that support the space industry.)

Impacts on the telecommunications sector. Being able to transfer and broadcast information worldwide instantaneously has been a revenue multiplier effect since the 1980s for telephone and television companies, and more recently Internet providers. A study by Euroconsult shows that the EUR 5 billion invested in the manufacturing and launch of telecom satellites in 2002 generated revenues of around EUR 100 billion in the largest telecommunications sector (Achache, 2006).

Impacts on the national economy. Some countries are trying to assess the possible impacts on the economy of investing in space systems. For example, in February 2006, the US Federal Aviation Administration's Office of Commercial Space Transportation (FAA/AST) published a report on the impacts of commercial space transportation and related industries on other industries. The report looked specifically at economic activity (revenues) and jobs throughout all industries in the national economy (FAA, 2006) using the economic impact analysis used an input/output method and the Regional Input-Output Modelling System (RIMS II) developed by the US Department of Commerce, Bureau of Economic Analysis. As defined by the FAA, the space sector was found to be responsible for USD 98 billion in economic activity in 2004 and 551 350 jobs throughout the United States via direct, indirect and induced impacts (see definition, tables and figures here). All major sectors are affected positively to some extent, including information services, manufacturing, finance and insurance, health care and social assistance. By comparison, using the same methodology, the economic impact of the civil aviation industry was found to be over 10 times that of commercial space transportation and enabled industries.

In terms of methodology, input-output analyses are valuable methods to measure economic impacts, as the FAA results show. One inevitable drawback in this type of analysis stems from the lack of precise space statistics. The statistical codes used cover, by definition, activities other than space. The original data for revenues, used to derive the economic impacts, are based on FAA adaptations of the private data of the Satellite Industry Association's (SIA) *2004 Satellite Industry Annual Indicators Study*. NAICS codes used in the FAA analysis include "334220: Radio and television broadcasting and wireless communications equipment manufacturing" for satellite manufacturing; "336414: Guided missile and space vehicle manufacturing" for launch vehicle manufacturing and services.

Local impacts. At the local level, economic spillovers can be felt in a given region due to the concentration of space-related activities. For example, with more than 1 600 NASA scientists and engineers, the direct global economic impact of the John C. Stennis Space Center (SSC) totalled USD 691 million in 2005, with a USD 503 million impact on the Mississippi and Louisiana communities within a 50-mile radius of the site. Likewise, impact studies show that onsite space activities at the "Centre Spatial Guyanais" (the European spaceport in French Guiana) represented 20% of this French department's GDP in 2005, with 1 350 people employed in the sector and 5 800 derived jobs in other sectors (one "direct" job being responsible for 4.4 "induced" jobs). In addition, actors involved in the space sector are responsible for 40% of local taxes and 60 % of French Guiana's imports (CNES and INSEE, 2005).

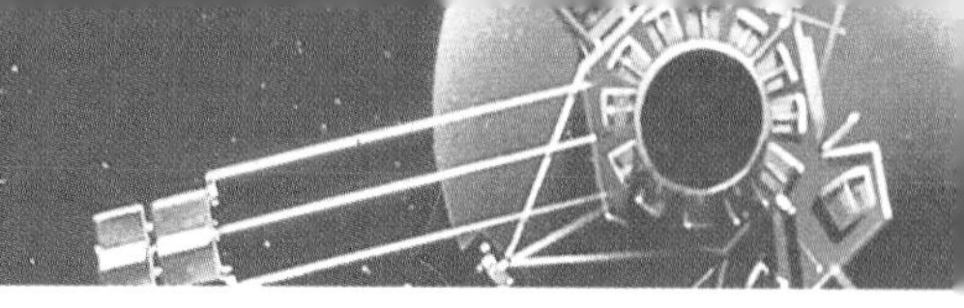

4.2. COMMERCIAL REVENUE MULTIPLIER EFFECT FOR NON-SPACE SECTORS

Table 4.2a. **Economic impacts of the US commercial space transportation and enabled industries, 2004 (thousands of USD)**

Industry	Direct impacts[1]	Indirect impacts[2]	Induced impacts[3]	Total
Launch vehicle manufacturing	286 936	759 171	612 277	1 658 384
Satellite manufacturing	626 307	1 654 746	1 185 058	3 466 111
Ground equipment manufacturing	5 722 370	15 118 905	10 827 507	316 68 782
Satellite services[4]	9 428 956	26 684 009	20 346 240	56 459 205
Remote sensing	69 529	279 196	332 474	681 199
Distribution industries[5]	532 049	1 886 862	1 734 366	4 153 278
Total Impacts	16 666 148	46 382 890	35 037 924	98 086 960

1. *Direct impacts* are the expenditures on inputs and labour involved in the provision of any final good or service relating to the industries analysed.
2. *Indirect impacts* involve the purchases (*e.g.* silicon, copper wire) made by and labour supplied by the industries that provide inputs to the launch and enabled industries. This type of impact quantifies the inter-industry trading and production necessary to provide the final goods and services.
3. *Induced impacts* are the successive rounds of increased household spending that result from the direct and indirect impacts (*e.g.* a launch vehicle engineer's increased spending on household goods and services).
4. Includes both end-user services (satellite telephony, VSAT services, satellite data services, and Direct to Home satellite, DTH) and transponder leasing (companies that operate satellites and lease or sell satellite transponder capacity.)
5. Includes wholesale and retail trade margins and transit costs (truck, air, and rail transportation services) incurred as components are moved to manufacturing sites.

Source: FAA (2006), *The Economic Impact of Commercial Space Transportation on the US Economy: 2004*, February

Figure 4.2. **Impacts of US commercial space transportation and enabled industries, 2004**

As a Percentage of all Impacts

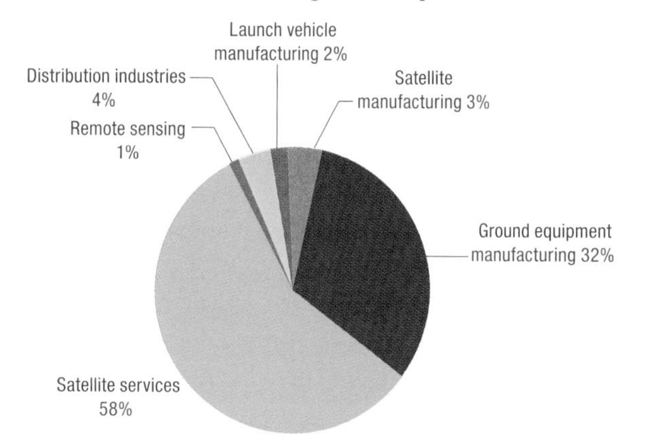

Source: FAA (2006), *The Economic Impact of Commercial Space Transportation on the US Economy: 2004*, February

Table 4.2b. **Economic impacts (revenues and jobs) throughout major US industry sectors, generated by commercial space transportation and enabled industries, 2004**

Thousands of US dollars and number of employees

Industry	Revenues ("economic activity")	Jobs
Information services	29 575 613	116 800
Manufacturing	27 439 628	87 820
Real estate and rental and leasing	6 571 523	15 250
Finance and insurance	4 776 096	22 600
Wholesale trade	4 686 286	28 830
Professional, scientific and technical services	4 159 086	34 260
Health care and social assistance	3 482 882	44 720
Retail trade	2 963 727	43 160
Transportation and warehousing	2 331 069	19 620
Other services	2 072 797	25 080
Accommodation and food services	1 777 420	31 540
Management of companies	1 761 363	12 080
Administrative and waste management services	1 600 600	27 300
Arts, entertainment and recreation	1 364 960	19 330
Utilities	1 292 394	2 850
Agriculture, forestry, fishing and hunting	881 254	5 840
Educational services	557 315	9 790
Mining	456 971	1 270
Construction	335 976	3 220
TOTAL	**USD 98 086 960**	**551 360 employees**

Source: FAA (2006), *The Economic Impact of Commercial Space Transportation on the US Economy: 2004*, February

4.3. IMPACTS ON KEY SOCIETAL CHALLENGES (ENVIRONMENT, NATURAL DISASTERS)

Thanks to their unique characteristics, space applications could make a considerable contribution to several long-term and enduring challenges of the 21st century: the environment, the use of natural resources, the management of natural disasters, international mobility, and the move to the knowledge society (OECD, 2005). This section provides brief examples of demonstrated benefits derived from existing space systems (see also Annex B, for some findings from a OECD Global Forum on Space Economics' case study on water resources management).

Today, a number of key activities could not operate without space systems. But like water, energy or other infrastructures, space systems and their products (*e.g.* communications links, imagery) have become so embedded in our modern societies that their benefits go largely unnoticed, except when systems fail to function as expected. Weather forecasting is a good example. Quasi-real-time space data integrated into better computer models have significantly improved the economic value of weather forecast information (e.g. precipitation forecasts for agriculture, temperature variations for electric utilities). The traditional error in a three-day forecast landfall position of hurricanes has been reduced from about 337 kilometres in 1985 to about 177 kilometres in 2004 (SSB, 2005), giving more time to warn populations and businesses.

The improved accuracy is partly due to better knowledge of the oceans gained from spaceborne observation. Ocean monitoring from space became operational in the late 1990s with the Franco-American mission Topex/Poséidon. Satellite altimetry enables water levels to be measured to within 3 cm at basin level, but also provides information for monitoring phenomena such as variations in ocean circulations like El Niño, seasonal changes in oceans, and tide mechanisms.

In most instances, space infrastructure needs to be closely combined with terrestrial facilities to become fully effective (*e.g.* Earth observation data need to be integrated in models with complementary non-space data). The main benefits from using space systems for tackling societal challenges generally take the form of: decreased transaction time, cost savings, cost avoidance, improved productivity, and increased efficiency for end-users of space assets (Box 4.3b). A number of cost-benefit studies on space systems have been conducted over the years, however it is still challenging to track specific benefits to space technologies. The calculation of the ratio between the costs of the system and the flow of the socio-economic benefits obtained is at times contentious because space solutions can rarely be considered in isolation. Diverse methodologies are used in benefits assessments, in particular non-market valuation methods (*e.g.* contingent valuation, choice modelling), due the public good nature of some of the benefits (see also Annex B).

In natural disasters management, the benefits from space infrastructure are clearer. Space-based observations allow the Earth to be seen as a dynamic integrated system of land, water, atmosphere, ice and biological processes, while satellite telecommunications allow worldwide connections. This is increasingly useful, as the number of disasters worldwide has already exceeded 300 a year since 1998, and there is a clear upward trend for both economic and insured losses related to natural disasters (OECD, 2006). Floods play a dominant role in both the number of disasters and the number of people affected, due to increasing population growth along coastlines (Figure 4.3a). Disasters also entail important economic losses globally, with floods once again the most important factor (Figure 4.3b). In 2005, economic losses from catastrophes amounted to more than USD 170 billion.

Losses due to flooding can be reduced by the creation of adequate flood defences (e.g. levees) and water management infrastructures (*e.g.* reservoirs), providing warnings to the threatened area and taking the appropriate emergency actions to fight floods (*e.g.* evacuation). Remote sensing from space provides data for the whole cycle of information for flood prevention, mitigation, pre-flood assessment, response (during the flood), recovery (post-flood), and weather newscasts (with 0-3 hour forecasts). For example, according to the French Civil Protection agency, regular satellite observations over the past five years have contributed to improving the control of water pumping efficiency on a large scale during floods. In 2001, it took three months for the cities and villages in the French department of the Somme to dry out. During flooding in the Arles area of the south of France in 2003, the use of timelier satellite imagery contributed to more appropriate siting of the Civil Protection's water pumps, thus taking "only" three weeks to get the city dry again.

4.3. IMPACTS ON KEY SOCIETAL CHALLENGES (ENVIRONMENT, NATURAL DISASTERS)

Figure 4.3a. **Number of people affected per disaster type**

Extreme temperature, drought, wild fires
Earthquake, volcano, landslides
Epidemic, famine
Flood
Wave, surge
Wind storm

Year
2005
2002
1999
1996
1993
1990

0 | 100 000 000 | 200 000 000 | 300 000 000 | 400 000 000 | 500 000 000 | 600 000 000 | 700 000 000

Source: OECD (2006), *Information Technology Outlook*, OECD, Paris.

Figure 4.3b. **Economic and insured losses due to disasters: Absolute values and long-term trends, 1950-2005**

Billions of constant 2005 US dollars

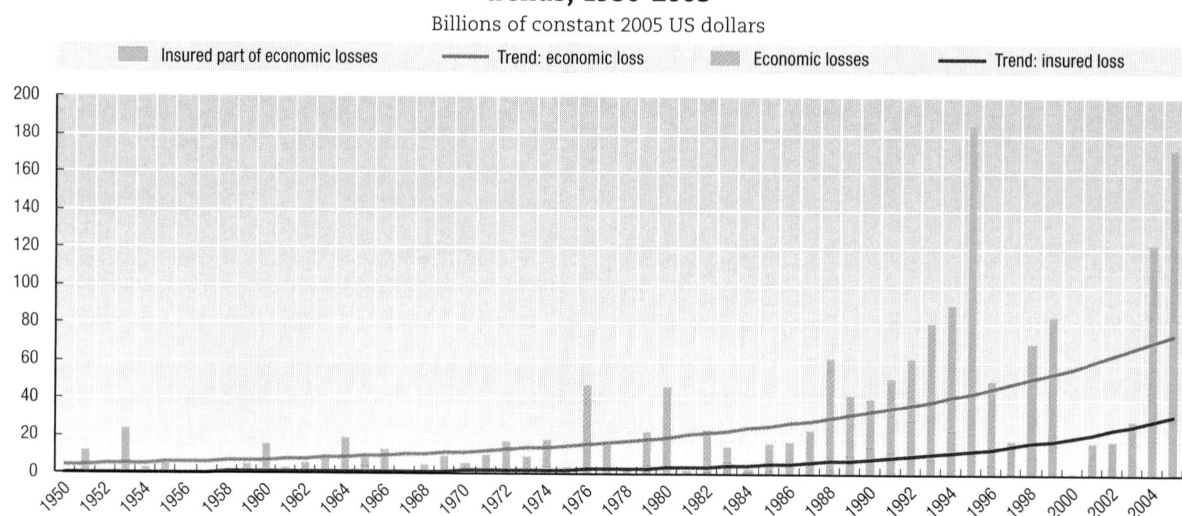

Source: OECD (2006), *Information Technology Outlook*, OECD, Paris.

Box 4.3. **Better efficiency due to the use of raw satellite data streams**

In the United Kingdom, the Met Office and other numerical weather prediction centres (NWP) conducted several impact studies in 2001 concerning the data from polar-orbiting satellites. This is a complex question since the NWP products can be affected as much by improvements in modelling techniques as by improvements in initial data.

Improvement of the meteorological forecast system The success of a meteorological forecast system can be measured in terms of accuracy through an index devised for the purpose. A Met Office internal review of the situation early in 1998 showed that a 3.5% improvement in the index could be attributed to the use of the raw satellite data stream from NOAA satellites. Further data and modelling improvements were introduced during 1999, which raised the index by about 5%. Impact studies have shown that about 2% can be attributed to better processing of NOAA data.

Uniqueness of data The NOAA imagery, although less frequent than that from the European Meteosat, has channels that, amongst other things, enable fog to be distinguished from low cloud. The altitude allows a resolution of 1km to be achieved below the satellite and the view in northern latitudes is not as foreshortened as it is from Meteosat. This raw imagery has become an essential forecasting aid, and has prompted European governments to develop their own specific polar orbit satellites (first European *Metop* satellite, launched in 2006), complementing NOAA's imagery.

Skills base: On the technical side, the development and provision of instruments has allowed the Met Office to develop a skills base that enables it to act as an intelligent customer for procuring 'best value for money' satellites through the various space programmes, and also gives it the credibility to influence the development of future programmes significantly for the benefit of the UK. What has been saved in terms of cost and improved performance through these activities is estimated to be many times the cost of retaining the team.

4.4. IMPACTS OF SPACE PROGRAMMES ON SPACE FIRMS

By definition, space programmes aim to support national or regional capabilities to develop space systems. The ability of firms to increase their space funding, secure new customers or create new activities has been studied by governments and space agencies.

As an example of recent findings by a small but active European Space Agency member, Norway has recently increased its involvement in space, partly due to a rather strong "spin-off effect" in its industries. In 2005, Norway found that for each million Norwegian kroner (NOK) of governmental support through ESA or nationa support programmes, the Norwegian space secto companies have on the average attained an additiona turnover of NOK 4.4 million (EUR 510 000). This spin-of effect factor of 4.4 is expected to climb further (Figure 4.4). Although this impact measure may vary widely depending on the country and the level of its specialisation (*e.g.* applications *versus* manufacturing), it is indicative of possible increased internationa competitiveness due to space involvement.

Figure 4.4. **Norway space industry spin-off factor, 1997-2005, with company forecasts up to 2009**

Actual and projected spin-off multiplier: non-ESA sales per euro of ESA/NSC contracts

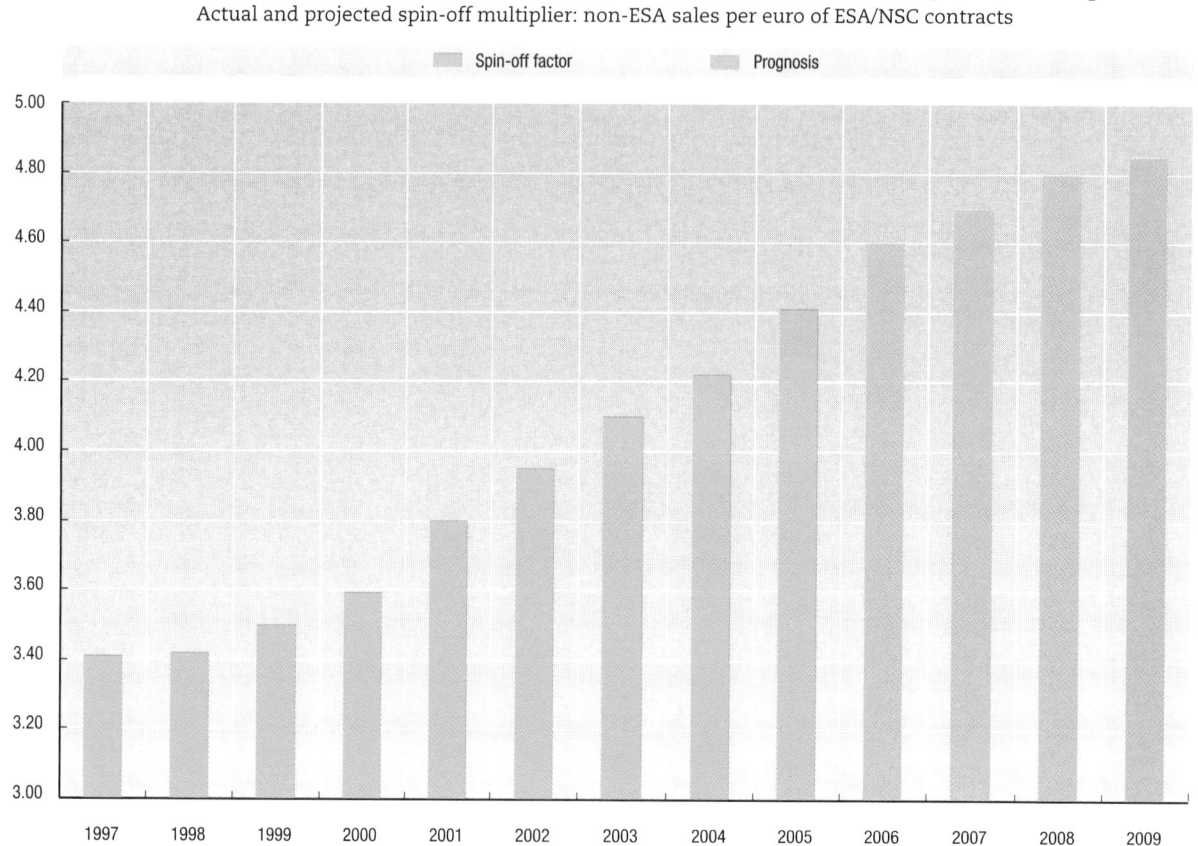

Source: Norwegian Space Centre (2005), *Annual Report*, Oslo.

4.5. THE WAY FORWARD

Improvements in the measurement of the space economy offer considerable potential for future analysis of its economic and societal impacts. Seeking better quality and comparability of economic data for the space sector and a better mapping of the downstream (often non-space) industry will be key steps in determining the small and large scale impacts of space activities.

Analysis of the information and communication technology (ICT) sector, which has been studied at the OECD for more than a decade, provides useful pointers for the space economy. Today, it is almost a given that ICT has impacts on the economic performance and success of individual firms, in particular when combined with investment in skills, organisational change and innovation (OECD, 2004). These impacts can be observed in firm-level studies for all OECD countries, but have not yet translated into better economic performance at the industry or economy-wide level in many OECD countries. Several factors with parallels in the space economy may explain this gap between firm-level and aggregate performance, *e.g.* ICT data measurement issues between countries, aggregation effects, time lags.

For space, international co-operation across national statistical offices (and other institutions) will be required.

Much work remains to be done to develop universal, data-driven indicators for the emerging space economy. More efforts in that direction could benefit both decision-makers, industry and citizens, and help them have a better understanding of the significance of space activities in the larger economy.

Further actions could include international efforts to separate the statistical classifications for aircraft and spacecraft industries, as well as exercises that drill down on space-related services (such as telecoms, satellite navigation). Case studies that assess the social and economic impacts of space applications in today's world would help to better qualify and quantify the space economy. The OECD Global Forum on Space Economics could be the platform that provides the impetus for such work, while further international co-operation will be required with national statistical offices, space agencies and industry associations.

Data sources

- Achache, Joseph (2006), *Les sentinelles de la Terre*, Hachette, Paris.
- Bach, L. *et al.* (2002), "Technological Transfers from the European Space Programme: A Dynamic View and A Comparison with Other R&D Projects", *Journal of Technology Transfer*, Vol. 27, No. 4, December.
- CNES and INSEE (2005), *Économie du spatial en Guyane*, *www.cnes.fr/web/3919-economie-du-spatial-en-guyane.php*, *accessed December 2006.*
- FAA (2006), The Economic Impact of Commercial Space Transportation on the US Economy: 2004, February.
- National Aeronautics and Space Administration (2006), *NASA Facts*, John C. Stennis Space Center, FS-2006-05-00027-SSC, May.
- OECD (2004), The Economic Impact of ICT: Measurement, Evidence and Implications, OECD, Paris.
- OECD (2005), Space 2030: Tackling Society's Challenges, OECD, Paris.
- OECD (2006), Information Technology Outlook, OECD, Paris.
- Space Studies Board (SSB) (2005), *Earth Science and Applications from Space: Urgent Needs and Opportunities to Serve the Nation*, Committee on Earth Science and Applications from Space, National Research Council, Washington DC.

5. SPOTLIGHTS ON SPACE ACTIVITIES OF SELECTED COUNTRIES

This chapter looks at space developments of some members of the OECD Global Forum on Space Economics. The countries covered are the United States, France, Italy, the United Kingdom, Canada, and Norway.

The data come from their own official sources (such as national space agencies or statistical offices), as well as private sources (in some cases). Direct comparisons between countries are not recommended due to definitional, conceptual and methodological differences. The particular characteristics or variables that are examined depend upon such factors as the type of available data, the timeliness of the data, reliability of the source and the ability of the data/variable to provide relevant insights to the reader.

5.1. UNITED STATES

Most of the emphasis in this section is on the US space-related manufacturing segment. Some attention is given to the space services sector by examining satellite telecommunications.

Highlights

Official US data on value added and employment for space manufacturing reveal that, while both have recovered somewhat from their lows earlier in this decade, they still continue to be well below their highs of the late 1990s, with 2004 value added of USD 10.6 billion and 64 000 employees (Figure 5.1a). The relative productivity of the workforce has also declined (Figure 5.1b), while space industry value added has fluctuated significantly (Figure 5.1c).

Satellite telecommunication revenues have shown a similar pattern to space industry revenues (Figure 5.1d). Although they seem to have recovered from their lows earlier in this decade, they are still well off their highs and accounted for only about 1.7% of total telecommunications revenue in 2004.

Definition

Official US statistics related to space have two main components: manufacturing and services. The manufacturing data come from the US Census Bureau's *Annual Survey of Manufactures* and encompass three industry groupings from the North American Industrial Classification System (NAICS): 336414 (Guided missiles and space vehicle manufacturing), 336415 (Guided missiles and space propulsion unit and propulsion unit parts manufacturing), and 336419 (Other guided missile and space vehicle parts and auxiliary equipment manufacturing). As it is not possible to separate the missiles from space vehicles, the two are together termed the US "space industry".

The space services sector data come from the US Census Bureau's Service Annual Survey data for NAICS 51334 (Satellite telecommunications). This industrial group refers to establishments that primarily engage in providing point-to-point telecommunications services to other establishments via a system of satellites or reselling of satellite telecommunications. Other areas also include satellite-based services (*e.g.* "Radio networks", "Cable networks"), but these were not included because space-based services and applications were believed to represent a very small portion of these large segments.

The North American Industry Classification System (NAICS) distinguishes between manufacturing of space equipment *versus* other aerospace as well as satellite communications. The main categories covering the space industry in the 2002 NAICS are:

The space industry: *a)* Class 336414: Guided missiles and space vehicles; *b)* Class 336415: Guided missiles and space propulsion unit and propulsion unit parts manufacturing; *c)* Class 336419: Other guided missile and space vehicle parts and auxiliary equipment manufacturing; *d)* Class 517410: Satellite telecommunications.

These aspects can thus be distinguished through official data. Other parts of the US space sector are more difficult to locate in official statistics. They include: *a)* Class 334220: Radio and television broadcasting and wireless communications equipment manufacturing. This class covers areas such as Global Positioning System (GPS) equipment, equipment for ground stations, satellite manufacturing, and satellite communications equipment. *b)* Class 334511: Search, navigation, etc., equipment. Covers navigational equipment. *c)* Class 515111: Radio networks. Includes satellite radio networks. *d)* Class 515210: Cable and other subscription programming. Includes satellite television networks. *e)* Class 517510: Cable and other programme distribution. Includes direct broadcast satellite, direct-to-home satellite systems, satellite distribution systems, etc. *f)* Class 517910: Other telecommunications. Includes satellite tracking and satellite telemetry.

Methodology

The data come from two US Census Bureau reports. The space manufacturing sector data come from the *Annual Survey of Manufactures*. These survey results are based on a sample of about 55 000 manufacturing establishments with one or more paid employees. As many manufacturing establishments also engage in non-manufacturing activities (captive services, such as payroll, R&D), the data split these services into the respective NAICS categories and not manufacturing. The report states that it is subject to sampling and non-sampling errors. Space services data are from the Bureau's *Services Annual Survey*.

Data comparability

The comparability of manufacturing data presented here is considered to be quite high, as the data were not affected by the revisions to the 2002 Census. Nevertheless, the concepts, definitions and methodology used by various government departments may not be the same as those of other data sources, so careful examination must be made when comparing data sources.

Data sources

- US Census Bureau (2005), *Annual Survey of Manufactures*, Washington DC, December.
- OECD (2006), *OECD Structural Analysis Statistics, STAN Industry*.

Figure 5.1a. **US space manufacturing industry employment and value added**

Employment in thousands, and value added in billions of US dollars

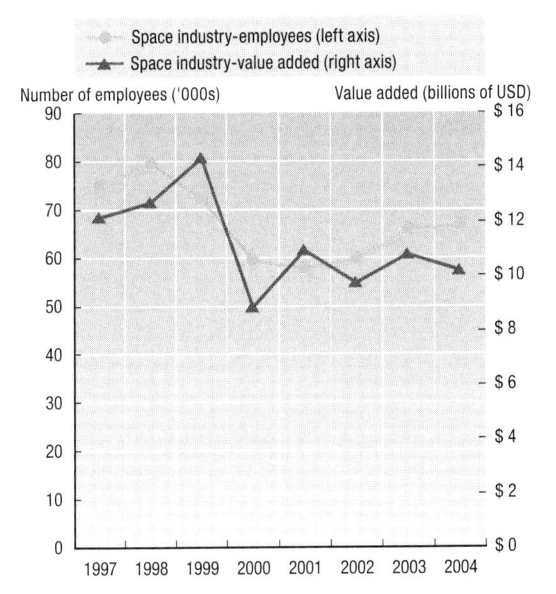

Figure 5.1b. **Contribution of space industry to US economy, 1997-2004**

Space industry as percentage of manufacturing employees and value added

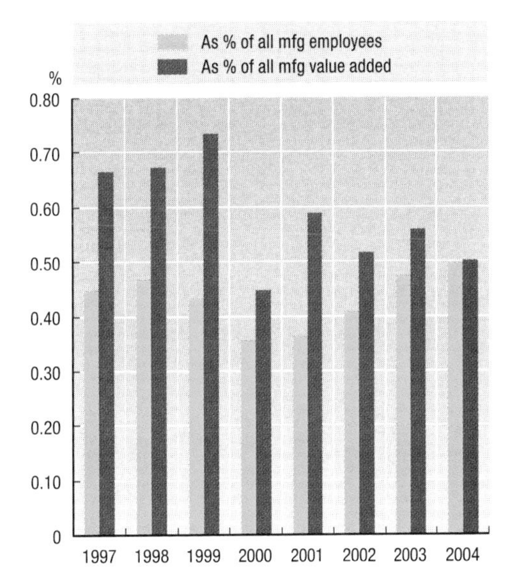

Source: US Census Bureau (2005), *Annual Survey of Manufactures*, Washington DC, December.

Figure 5.1c. **Space *versus* manufacturing value added growth rate, 1998-2004**

Percentage of total manufacturing sector value added

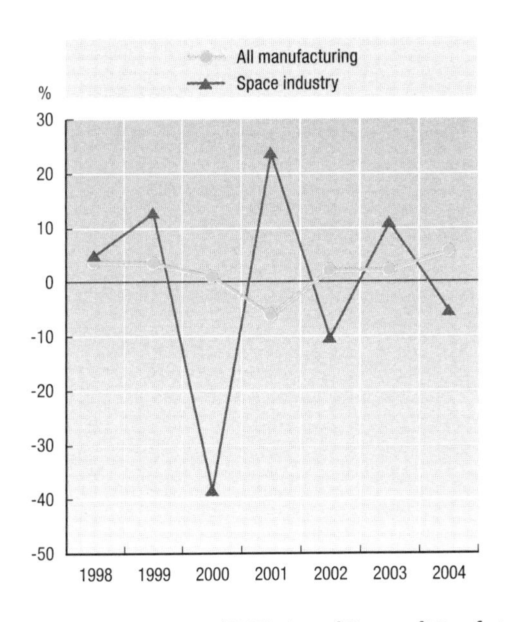

Source: US Census Bureau (2005), *Annual Survey of Manufactures*, December.

Figure 5.1d. **US satellite telecom revenue and percentage of telecom revenue, 1998-2004**

Satellite revenue in billions of US dollars and as percentage of total US telecom industry revenue

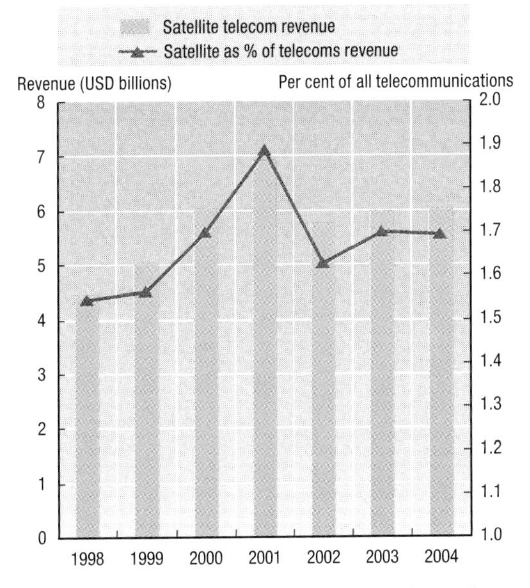

Source: US Census Bureau (2005), *Annual Survey of Manufactures* Washington DC, December.

5.2. FRANCE

France has had a national budget dedicated specifically to space activities since the 1960s. It is the largest budget for space activities in Europe.

Highlights

Manufacturing turnover in the French space sector is estimated at around EUR 2.7 billion by the Service for Industrial Studies and Statistics (SESSI) in official statistics (Table 5.2a) and more than EUR 3 billion in 2005 by GIFAS, the French Aerospace Industries Association (unconsolidated data) (Figure 5.2a, Table 5.2b).

Although limited in scope, National Institute for Statistics and Economic Studies (INSEE) surveys provide an interesting view of how space activities affect a large number of firms, and not only the "big actors". In the Midi-Pyrenées region (especially Toulouse), 221 of 558 INSEE survey respondents declared an activity linked to space in 2004 (190 of these 221 also work with the aeronautical sector). For these firms, sales related to space activities represented 14% of revenue. Companies active in electrical and electronic equipment production, computers, and service providers are the most dependent on space contracts (Figures 5.2b and 5.2c). In French Guiana, in 2003, 64 subcontractors and service providers from different sectors (industry, construction, energy), employing 2 100 people, reported 48% of turnover relying on space-related contracts (INSEE Guiana survey, 2005).

As elsewhere, in France telecommunications is the strongest generator of space applications revenues (61%), followed by Earth observation, scientific instruments, and navigation (Figure 5.2d).

Around 11 500 of the 30 000 Europeans who work in the space industry work in France (GIFAS, 2006), but this might be underestimated as the space services sector is not well identified in official or private data.

Definition

As in other countries, space statistics in France are often "lost" in aggregated aerospace figures. However, the French statistical classification system (NAF) provides a notable level of detail for space activities, which could be useful if these categories were used more widely in industry surveys (see Box 5.2). The official data presented here are based on regional surveys by INSEE, and provide details on the French space manufacturing and services sector, via an analysis of their contractual dependence (Figure 4.2). Further information is provided by SESSI which conducts annual surveys on specific industrial branches. For the space sector, it looks in particular at NAF category 35.3C "*Manufacturing of launchers and space vehicles*", although this category excludes missiles, launch services provision and satellite operations. For data on French space application activities (telecommunications, remote sensing, etc.), data from GIFAS are included.

Methodology

Official French statistics tend to focus on manufacturing more than services, which need closer examination. The SESSI provides a rigorous annual survey of the main space manufacturing companies active in the sector (seven in 2005). The surveys are limited to current classification systems and exclude many other actors. Data from INSEE are complementary. INSEE conducts regional surveys in the Midi-Pyrenées region (annually since 1982), Aquitaine (annually since 2000), and French Guiana (regularly, but not annually), specifically covering manufacturers, subcontractors and service providers in the aeronautical and space sectors. The surveys provide snapshots of the French aerospace industry, an important sector for the economies of the three regions in terms of revenues and employment.

Data comparability

The data presented here come from different organisations and perspectives, and even when looking solely at the official data, may not be comparable over time. The INSEE surveys provide interesting regional snapshots of the industry, although according to INSEE, the data collected yearly are not fully comparable with previous years, due to the variation in the response rates from year to year. This is normal for surveys of this type, and due to the specificities of the regional responses, the results cannot be generalised. The SESSI data offer long-term generated statistics, although limited in scope. The GIFAS data are based on private industry surveys of its members, with different methodologies used over time, due to this evolving and consolidating industry. This also influences the comparability of data over the years.

Data sources

- GIFAS (2006), Rapport Annuel 2005-2006, Présentation des principaux agrégats concernant les sociétés du GIFAS, Paris.
- INSEE Aquitaine (2005), Aéronautique-Espace : résultats de l'enquête 2005, Dossier n°56, December.
- INSEE Midi-Pyrénées (2005), Enquête aéronautique, espace et sous-traitance, Dossier n°132, December.
- INSEE Guyane (2005), Espace et sous-traitance : résultats de l'enquête de sous-traitance 2003, December.
- SESSI (2006), Enquêtes annuelles de branche: industries aéronautiques 2005, Paris, 15 June.

THE SPACE ECONOMY AT A GLANCE – ISBN 978-92-64-03109-8 – © OECD 2007

Box 5.2. The space sector in French official statistics

France bases its statistical classifications on the General Industrial Classification of Economic Activities within the European Communities (also called NACE, Nomenclature d'Activité dans la Communauté Européenne). The main NACE categories are then broken down further in the French national statistical system NAF (Nomenclature d'Activités Française).

Category "353" in the NACE system covers "Manufacture of aircraft and spacecraft" (Construction aéronautique et spatiale). The French NAF provides a more detailed category with the 35.3C code: Manufacturing of launchers and space vehicles (Construction de lanceurs et engins spatiaux).

It would be useful if the following product subcategories were used more in industry surveys:

- Product 00020: Manufacturing of Ariane 5 launchers (Construction de lanceurs Ariane 5);
- Product 00021: Parts of launchers, including boosters (Parties de Lanceurs, y compris booster);
- Product 00022: Other launchers aside from Ariane 5, *e.g.* M51, M52, etc. (Autres lanceurs qu'Ariane 5 : M51, M52, etc.);
- Product 00030: Manufacturing of space vehicles (Construction d'engins spatiaux);
- Product 000301: Space vehicles and satellites (Engins spatiaux et satellites);
- Product 000302: Parts of space vehicles and satellites (Parties d'engins spatiaux et satellites).

Table 5.2a. Turnover from manufacturing of launchers and space vehicles in France in 2005 (NAF code: 35.3C[1])

Products (including exports)	SESSI Code	Number of companies	Turnover (in euros)
Space vehicles (launchers)	40001	6	1 037 436
Satellites	40002	2	N/A
Parts of satellites	40003	2	N/A
Total (extrapolated)		7	2 765 406

1. Missiles, launch services provision and satellites operations are excluded. N/A: Figures not publicly available.

Source: SESSI (2006), Enquêtes annuelles de branche: industries aéronautiques 2005, 15 June.

Table 5.2b. Evolution of French space manufacturing turnover, per activity and total

Unconsolidated, in millions of euros before adjustment for inflation

	2003	2004	2005	Evolution 2004-05
Space systems manufacturers	2 336	2 480	2 521	1.67%
Propulsion systems manufacturers	283	320	333	3.94%
Equipment manufacturers	116	196	189	–3.52%
TOTAL	2 735	2 996	3 043	1.57%
		Civil turnover	**Military turnover**	**Total Space turnover**
		2 556	487	**3 043**

Source: GIFAS (2006), Rapport Annuel 2005-2006, Paris.

5.2. FRANCE

Figure 5.2a. **Evolution of the French space manufacturing turnover by type of activities unconsolidated**

Millions of euros before adjustment for inflation

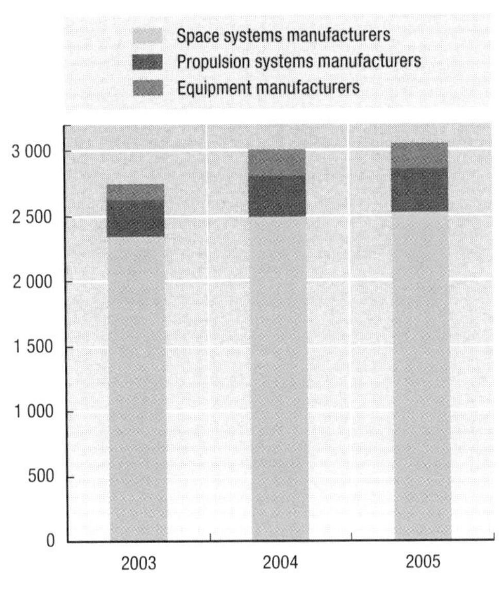

Source: GIFAS (2006), *Rapport Annuel 2005-2006.*

Figure 5.2b. **Space and aeronautics as Percentage of Turnover of 221 active firms in the aerospace sector in the Midi Pyrenees region, 2004 (%)**[1]

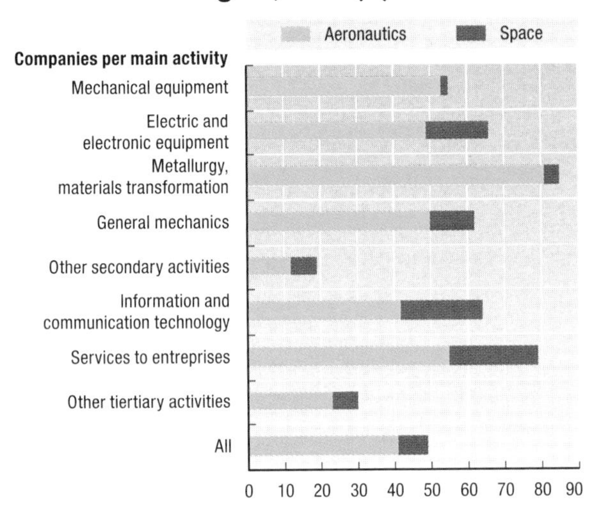

1. The data exclude revenues other than space and aeronautics.

Source: INSEE Midi Pyrenees (2005), *Enquête aéronautique, espace et sous-traitance.*

Figure 5.2c. **Space and aeronautics shares in aerospace firms turnover in Midi-Pyrénées and Aquitaine, 2004**

As a Percentage of all Aerospace Turnover

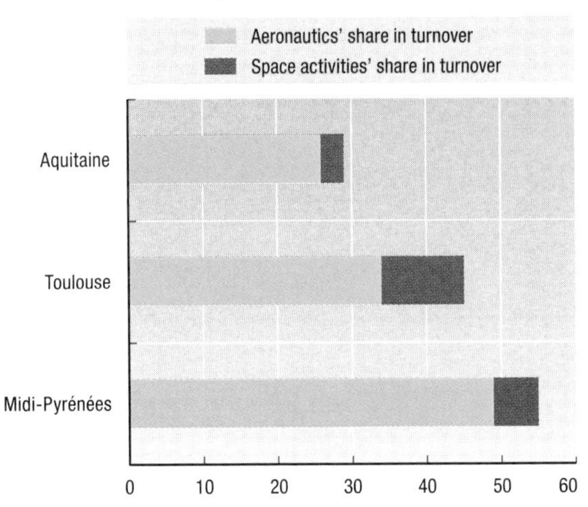

Source: INSEE Aquitaine (2005), INSEE Midi Pyrenees (2005).

Figure 5.2d. **Revenues for satellites and related space systems by applications**

As a percentage of total revenues

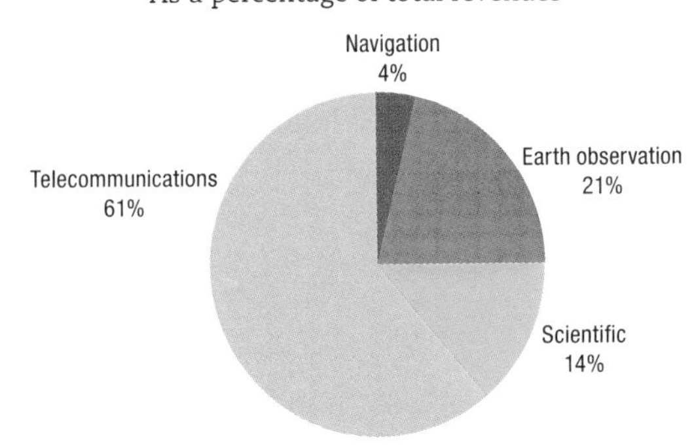

Source: GIFAS (2006), *Rapport Annuel 2005-2006.*, Paris.

5.3. ITALY

Italy is a key player in the space industry with over 180 enterprises, resulting in turnover of EUR 1.5 billion in 2005. This section provides a general overview of the national space sector by incorporating data from the Italian space agency, Agenzia Spaziale Italiana (ASI), and national industry associations.

Highlights

Of the 180 enterprises identified by ASI as being part of the Italian space industry, over half were involved in the services sector in 2005 (Figure 5.3a).

In terms of activity, Italian companies were mainly involved in production and systems integration (59%) and R&D and design (23%) activities (Figure 5.3b). Italy has an international customer base, and examination of ESA contracts awarded to Italy reveals that the largest portion was in the areas of launchers and human spaceflight (both at 28%), with Earth observation and science also significant (Figure 5.3c).

According to Italian industry association data, the space industry employed 6 220 personnel in 2005, with two-thirds employed in the manufacturing sector (Figure 5.3d). Total turnover that year was EUR 1.5 billion, with two-thirds in the space manufacturing industry (Figure 5.3e).

Definition

The Italian space agency (ASI) incorporates an interesting and comprehensive overview of the industry in its reports. The reports include the size and characteristics of the Italian space industrial sector and its national competencies, as well as other important players such as universities and research centres (*e.g. National Aerospace Plan 2006-2008*).

Other data sources used here include Italian industry associations: the Italian Industry Association for Aerospace Systems and Defence (AIAD); the Association for Space-based Applications and Services (ASAS); and the Association of Italian small and medium aerospace enterprises (AIPAS), which provides data related only to manufacturing and service industries.

Methodology

ASI provides annual data on universities, industries and research centres that receive contracts, including their role and competencies. The data from the industry associations incorporate information from surveys of their members (manufacturing and service companies). This encompasses turnover for each company, covering their contracts not only from ASI but also from other customers.

Data comparability

As in the case of other countries, the data presented here come from different organisations, with specific ways of conducting their industry surveys. This needs to be taken account when trying to make comparisons with other data sources.

Data sources

- Agenzia Spaziale Italiana (ASI) (2006), *National Aerospace Plan 2006-2008*, Roma.
- Agenzia Spaziale Italiana (ASI) (2006), *The Italian Space Context*, OECD Global Space Forum's Workshop in Roma, Italy, October.
- Italian Industry Association for Aerospace Systems and Defence (AIAD), Association of Space-based Applications and Services (ASAS), Association of Italian Small and Medium Aerospace Enterprises (AIPAS) (2006).

5.3. ITALY

Figure 5.3a. **Breakdown of Italian space enterprises by sector**

Percentage of total enterprises

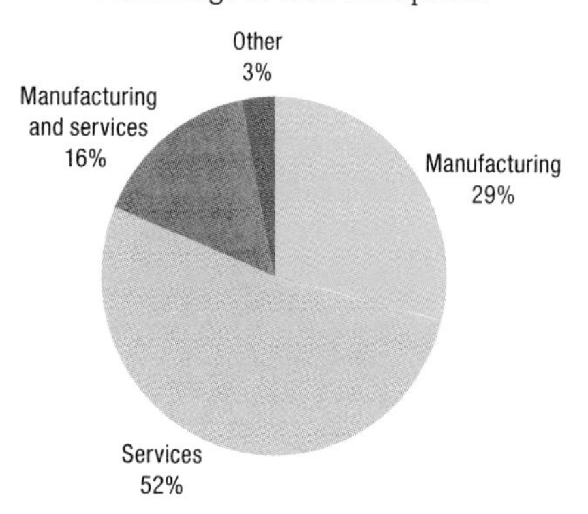

Figure 5.3b. **Breakdown of Italian space enterprises by activity/skill type**

Percentage of total enterprises

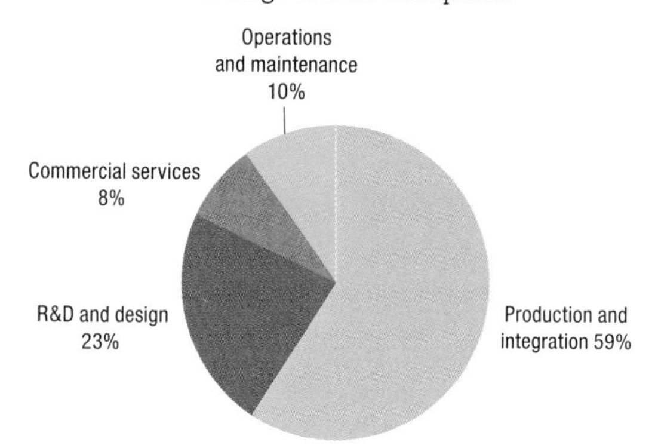

Source: Agenzia Spaziale Italiana (ASI) (2006), *The Italian Space Context*, OECD Global Space Forum's Workshop in Roma, Italy, October.

Figure 5.3c. **ESA contracts to Italy per directorate**

Percentage of all ESA contracts to Italy

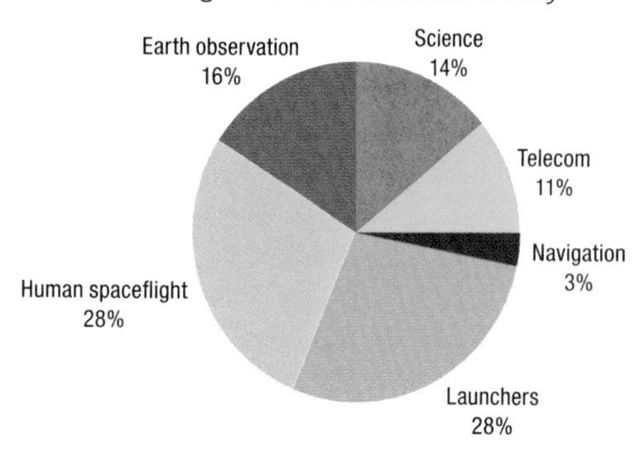

Source: Agenzia Spaziale Italiana (ASI) (2006), *The Italian Space Context*, OECD Global Space Forum's Workshop in Roma, Italy, October.

Figure 5.3d. **Employment in Italian Space Industry by industry type, 2005**

Actual number of Employees and as Percentage of industry Total

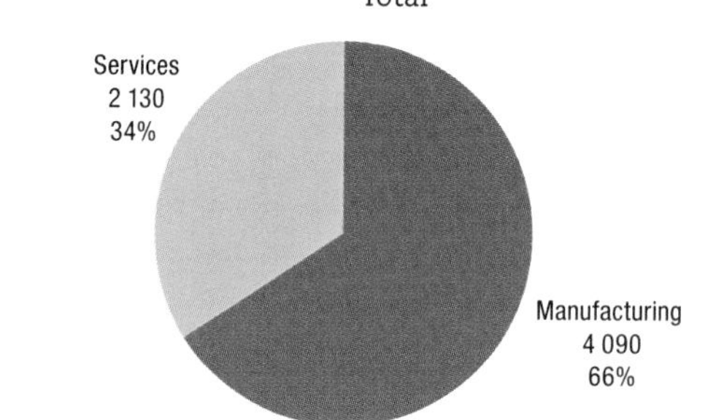

Source: Industry associations (AIAD, ASAS, AIPAS) (2006).

Figure 5.3e. **Employment in Italian Space Industry by industry type, 2005**

Actual number of Employees and as Percentage of industry Total

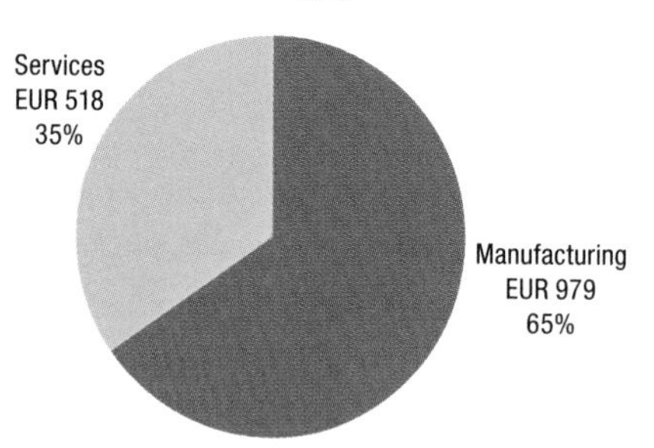

Source: Industry associations (AIAD, ASAS, AIPAS) (2006).

5.4. UNITED KINGDOM

The United Kingdom (UK) space industry is an important actor in Europe. This section provides an overview of the industry by examining both its upstream (*i.e.* space technologies and systems) and downstream applications (*i.e.* services that exploit the technology).

Highlights

An examination of UK space sector turnover shows that downstream activities have consistently provided about 85% of the GBP 4.3 billion total from 1999/2000 to 2004/2005 compared to 15% for upstream (Figure 5.4a).

In 2004/2005, the upstream market was dominated by the space prime market (*e.g.* large satellite manufacturing,) and satellite broadcast services overwhelmingly dominated the downstream category with 72% of the total (Figures 5.4b and 5.4c).

In terms of the regional location of UK space sector customers, in 2004/2005, 73% were in the UK, with another 14% accounted for by the EU (except the UK). The type of business sought by UK space clients was overwhelmingly in the consumer market, while non-UK regions were dominated by business commercial markets (Figure 5.4d). An examination of the uses to which these products were put in 2004/2005 reveals that satellite telecommunications accounted for 92% of total turnover (72% broadcasting and 20% communications) (Figure 5.4e).

Definition

In terms of upstream business activities suppliers, this category includes space prime, space subsystem suppliers, space component suppliers, contract R&D, and the ground segment. Downstream business activity categories were satellite broadcast services, satellite communication services, user equipment, financial services and other/support services.

"Applications" examines the use to which products in the activity areas above are ultimately put. These are broadcasting, telecommunications, Earth observation, space science, navigation, space transportation and other.

Methodology

The data are provided by the British National Space Centre (BNSC) based on a study performed on its behalf by Bramshill Consultancy Ltd. The study incorporates an examination of over 225 companies that are either totally or partially involved in the space sector (*i.e.* either upstream, providing space technology or downstream, using space technology) to give a picture of the "complete space sector". The data are often corrected to include the effects of inflation and non-response by companies.

Data comparability

Data comparability for the period examined by this report (1999 to 2005) tends to be quite consistent, with the exception of business categories that have been "redefined/extended", and when attribution between prime and subsystem suppliers has been "rationalised" compared to previous reports. Comparisons with other data sources, such as Eurospace, are not possible because Eurospace examines manufacturing units while BNSC data incorporate both manufacturing and services components.

Data sources

- British National Space Centre (BNSC) (2006), *Size and Health of the Space Industry Annual Report*, London.

5.4. UNITED KINGDOM

Figure 5.4a. **UK space industry upstream and downstream real turnover, 1999-2005**

Billions of British pounds

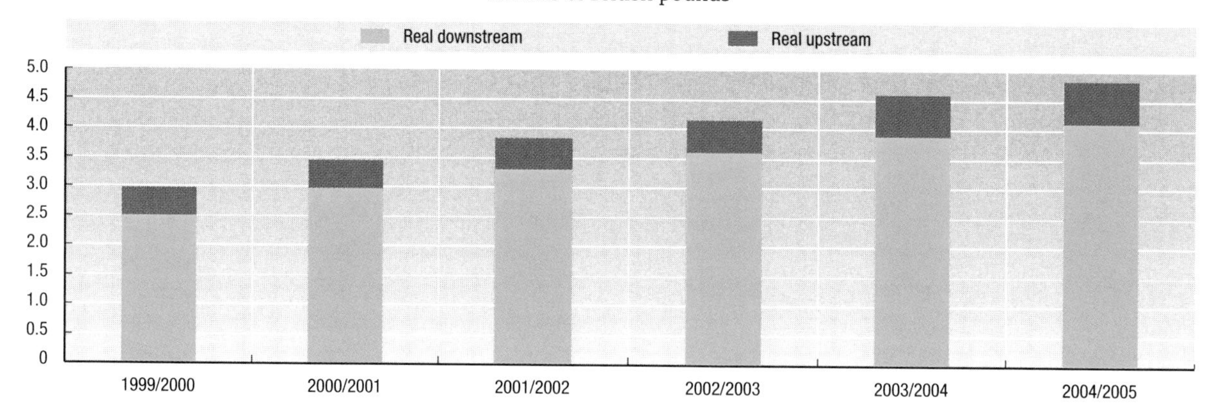

Source: BNSC (2006), *Size and Health of the Space Industry Annual Report*, London.

Figure 5.4b. **Breakdown of UK upstream turnover, 2004/05**

As a percentage of total upstream turnover

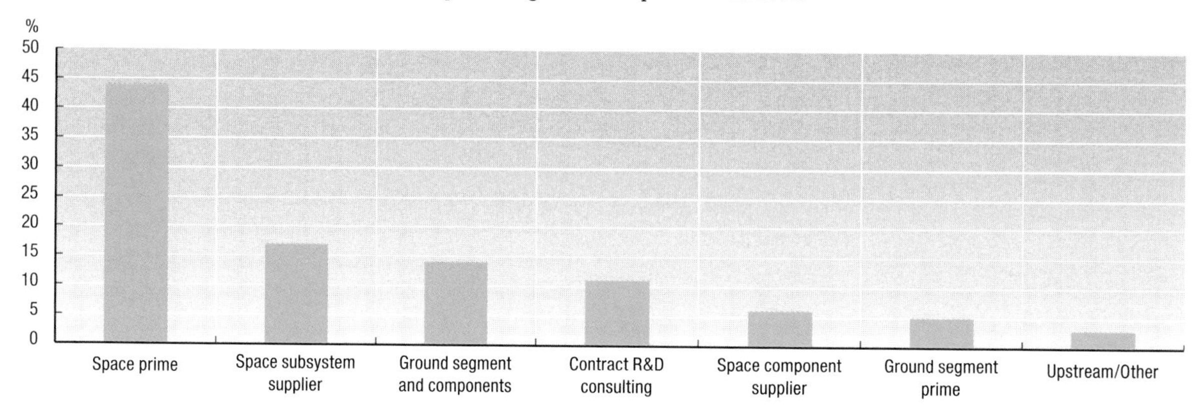

Source: BNSC (2006), *Size and Health of the Space Industry Annual Report*, London.

Figure 5.4c. **Breakdown of UK downstream turnover, 2004/05**

As a percentage of total downstream turnover

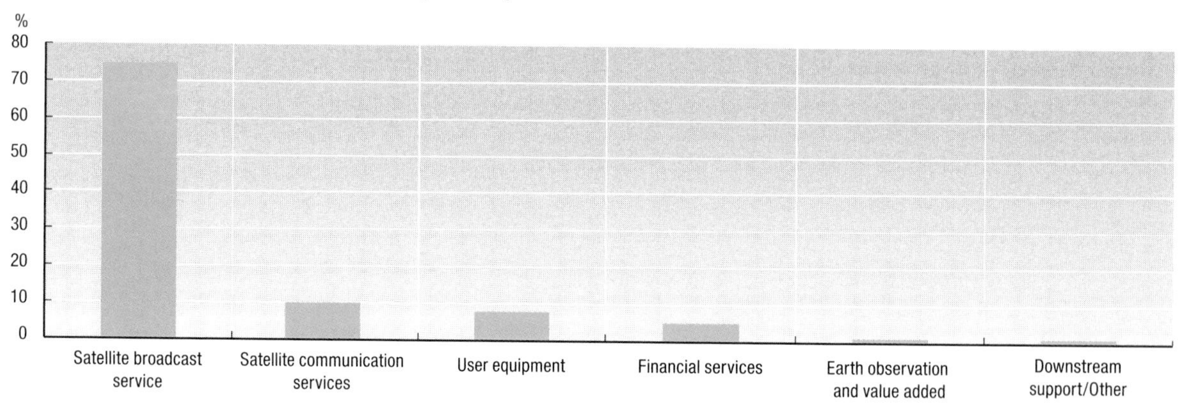

Source: BNSC (2006), Size and Health of the Space Industry Annual Report, London.

Figure 5.4d. **Turnover of UK space industry customers by region and type, 2004/05**

Billion of British pounds (Total: GBP 4.83 billion)

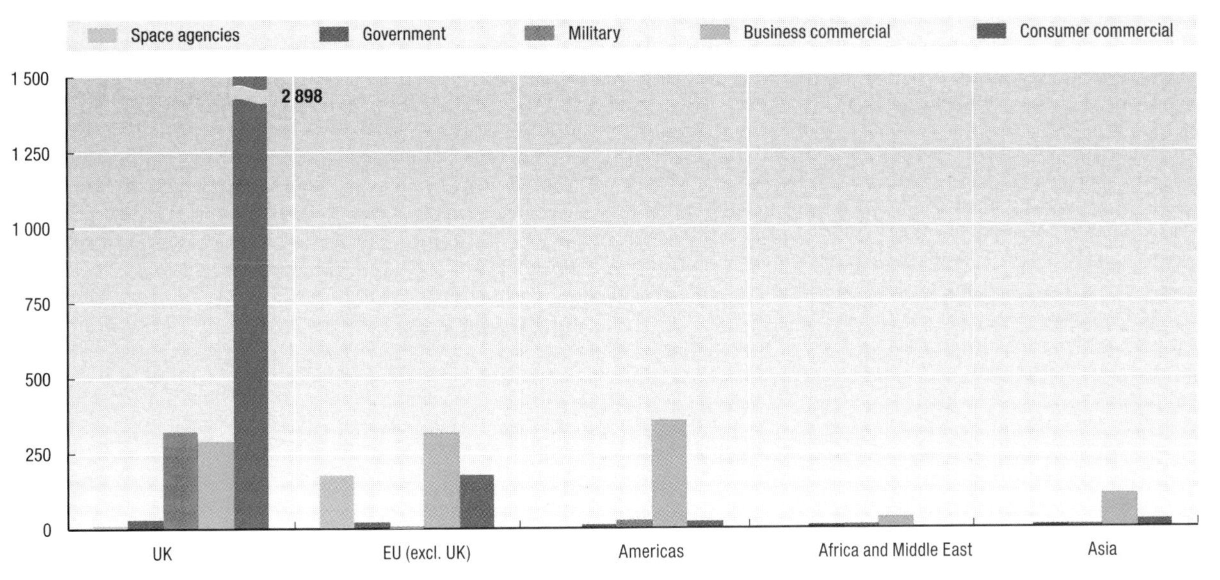

Source: BNSC (2006), *Size and Health of the Space Industry Annual Report*, London.

Figure 5.4e. **Breakdown of UK turnover by application, 2004/05**

As a percentage of total turnover

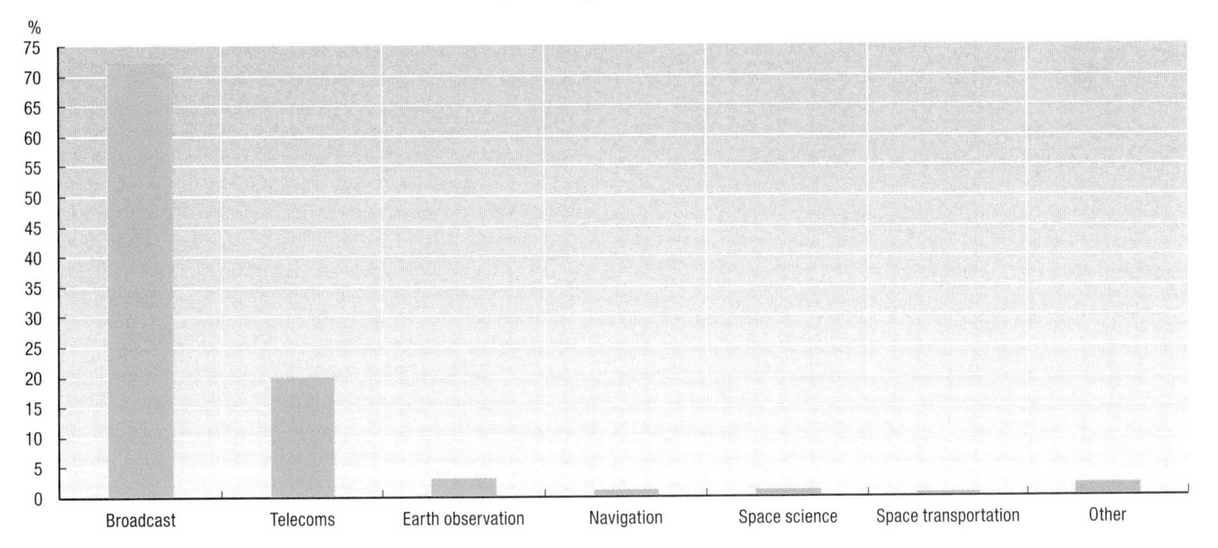

Source: BNSC (2006), *Size and Health of the Space Industry Annual Report*, London.

5.5. CANADA

Canada is particularly active in some space-related activities. The data presented here come primarily from the Canadian Space Agency (CSA).

Highlights

The Canadian space sector saw a slowdown in revenue growth in 2005 (up only 2%) compared with 11% in 2003 and 22% in 2004 (Figure 5.5a). Similarly, employment declined by 7% in 2005 from the previous year.

An examination of revenue breakdown shows that exports matched domestic revenue by accounting for 50% of the total in 2005 (Figure 5.5b). This shows a very significant rise from the 31% that exports accounted for in 1996.

Total domestic revenues have increasingly come from private sources (Figure 5.5c). In fact, public funds in 2005 of CAD 236 million were almost exactly the same as in 1996, while private sources more than doubled to over CAD 1 billion.

Most export revenue continued to come from the US throughout 1996-2004 (Figure 5.5d). However, this amount has fallen to less than 50% in recent years as Europe and "Other" (Oceania, South America and Africa, etc.) account for greater amounts. The biggest sources of revenue are applications and services (*i.e.* development and provision of services/products derived from space systems) (Figure 5.5e). Satellite communications constitute the largest revenue-generating sector, with 78% of total revenues in 2005 (Figure 5.5f).

Definition

The Canadian space sector refers to private, public and academic organisations that engage in activities that rely on the development and use of space assets and/or space data. The sector data are broken down into four specific types of activities: (1) the space segment (including R&D, manufacturing, testing and integration of systems and components); (2) the ground segment (facilities on Earth related to controlling space based systems and satellites); (3) applications and services (development of products and services using space systems/services); and (4) space research (primary R&D related to non- or pre-commercial space activities). These activities can also be sub-divided into sectors of activity: satellite communications; Earth observation; robotics; space science; navigation and other.

Methodology

The CSA sent surveys to over 200 organisations (including private entities, research organisations and universities) deemed to have a defined strategic interest in the space industry. The data are supplemented by international consultations with the CSA and relevant government officials that deal with stakeholders. The data are sometimes highly aggregated to protect the confidentiality of respondents.

Data comparability

The CSA indicates a margin of error of about 2.5%. The data – including its underlying assumptions, concepts and definitions – are believed to be generally internally consistent. However, there are a few differences (*e.g.* "Satellite Communications" was named "Telecom" in older reports). Also, note that "Navigation" was included under "Other" until 1998.

Data sources

- Canadian Space Agency (2000), *State of the Canadian Space Sector*, annual report.
- Canadian Space Agency (2005), *State of the Canadian Space Sector*, annual report.

5.5. CANADA

Figure 5.5a. **Canadian space sector revenues and employment, 1996-2005**

Revenues in billions of Canadian dollars and employment in employee numbers

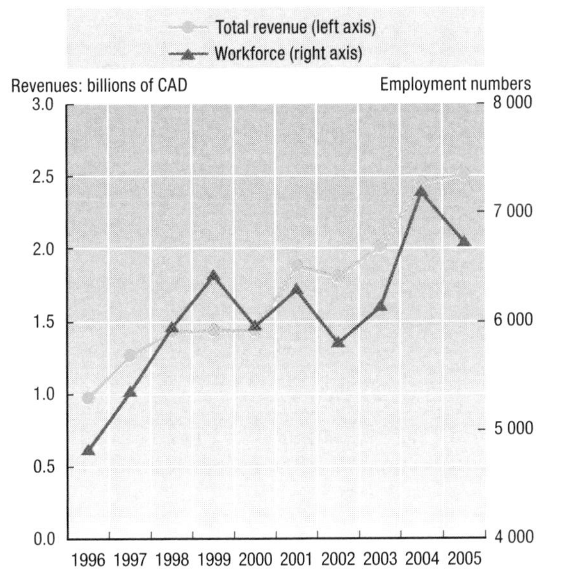

Source: Canadian Space Agency (2000, 2005), *State of the Canadian Space Sector*, annual reports.

Figure 5.5b. **Canadian space sector revenue breakdown, 1996-2005**

Billions of Canadian dollars

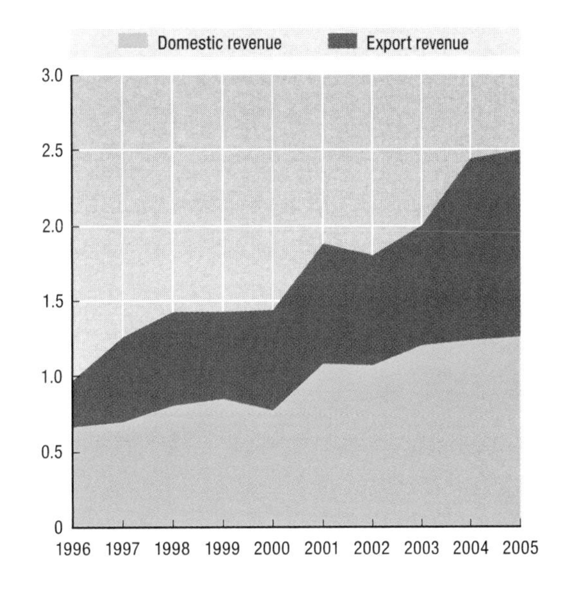

Source: Canadian Space Agency (2000, 2005), *State of the Canadian Space Sector*, annual reports.

Figure 5.5c. **Canadian space sector domestic revenue breakdown, 1996-2005**

Billions of Canadian dollars

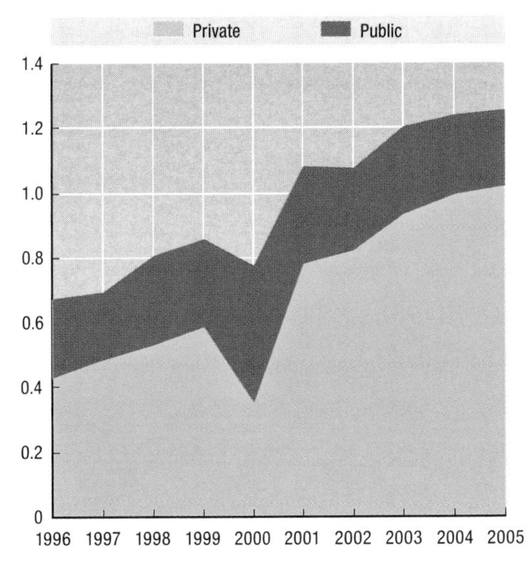

Source: Canadian Space Agency (2000, 2005), *State of the Canadian Space Sector*, annual reports.

Figure 5.5d. **Canadian space sector export revenue source breakdown, 1996-2005**

Billions of Canadian dollars

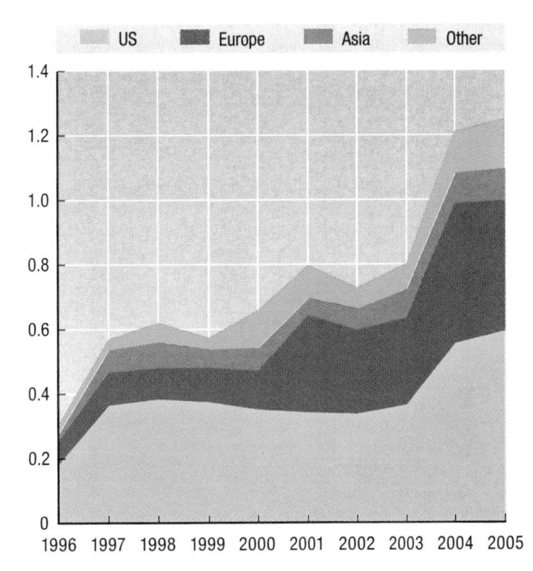

Source: Canadian Space Agency (2000, 2005), *State of the Canadian Space Sector*, annual reports.

5.5. CANADA

Figure 5.5e. **Canadian space sector total revenue by categories, 1996-2005**

Billions of Canadian dollars

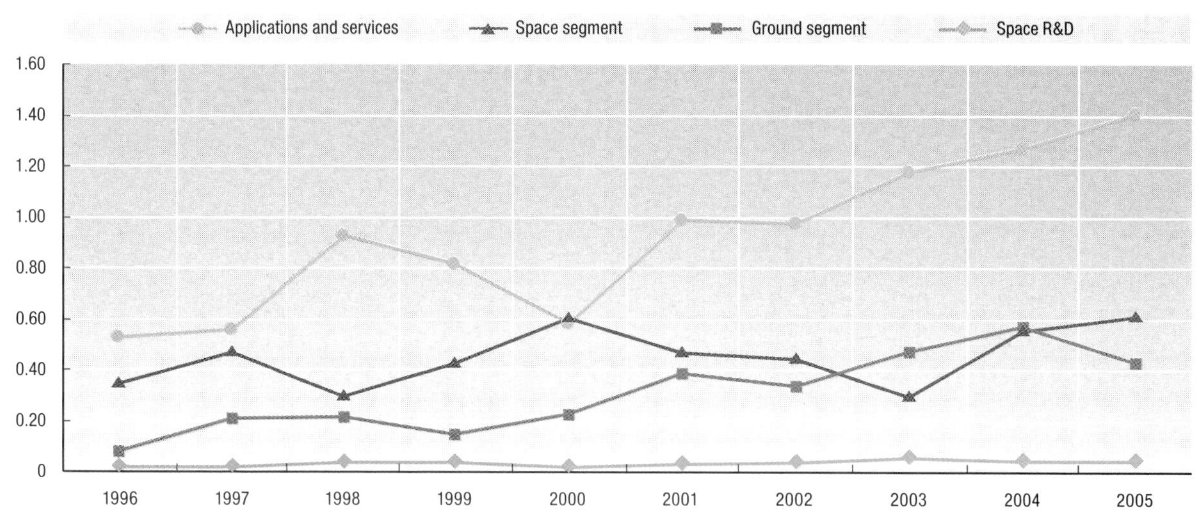

Source: Canadian Space Agency (2000, 2005), *State of the Canadian Space Sector*, annual reports.

Figure 5.5f. **Canadian space sector revenue by activity sector, 1996-2005**

Billions of Canadian dollars

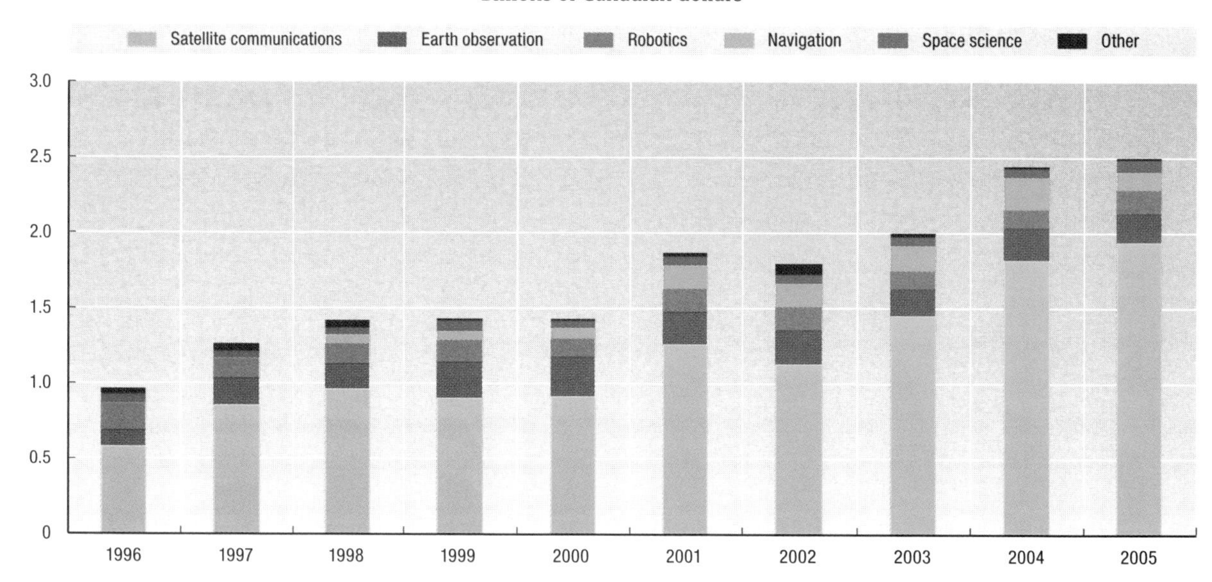

Source: Canadian Space Agency (2000, 2005), *State of the Canadian Space Sector*, annual reports.

5.6. NORWAY

While the Norwegian space sector may not be as large that of some of the other countries profiled here, Norway considers the sector essential in meeting its environmental, security, telecommunications, economic and research needs. The establishment of the Norwegian Space Centre (NSC) in 1987 set the stage for Norway to become an active player in the space sector.

Highlights

Because of its geography and its specific national requirements, Norway is pursuing several niche markets in the space sector (*e.g.* satellite telecommunications applications for its merchant fleet, oil and natural gas installations, and the Svalbard archipelago; radar satellite services for monitoring Norwegian waters).

Although turnover for space-related goods and services by Norwegian-based producers fell for the second consecutive year in 2005 to 5.2 billion Norwegian kroner (NOK) (Figure 5.6a), much of the decline was attributed to the rise of the kroner against the US dollar rather than to volume differences. Turnover is expected to rise for the next few years, although this depends upon strong private and public support. Exports accounted for 82% of Norwegian space-related turnover in 2005 (Figure 5.6b).

The "spin-off effect" indicates that the non-ESA sales impact of government support for ESA or NSC contracts has been consistently rising over the past nine years to a coefficient of 4.4 in 2005, with the trend forecast to continue into 2009 (Figure 5.6c). In 2005, the impact of this factor is reflected in the approximately EUR 140 million of these contracts resulting in EUR 616 million additional sales for Norwegian space sector companies (Figure 5.6d).

Norway closely examines the "spin-off effects" (or multiplier), which indicate how much the impact of one euro's worth of development contract via ESA or the NSC will amount to in additional non-ESA sales for Norway's space sector. The multiplier effect reflects how the technological advancements, development of new products and greater visibility provided via the ESA/NSC contracts translate into significant non-ESA sales.

Definition

Norwegian data on turnover for the space sector refer to goods and services offered by Norwegian-based companies and institutions (including research institutions). The statistics include data on both public and private customers including those contracts involving the ESA.

Methodology

In 2005, there were 21 companies specifically identified as being in the Norwegian space sector. Nevertheless, there are other institutions and companies (outside the space sector) that also play a significant role in the production of space-related goods or provision of services, and they also provided data.

Estimates of future values (for years 2006 to 2009) are forecast with assumptions of a rise in both public and private contributions to space. The "spin-off effects" factor, indicating the coefficient by which ESA and NSC contracts lead to further sales by space-companies, is obtained by carefully examining the relation between ESA/NSC contract activity and future business activities using a three-year time lag.

Data comparability

Concepts, definitions and methodology appear to be quite consistent throughout the various NSC annual reports. Nevertheless, it is important to keep in mind that the values are in Norwegian kroner (NOK). As such, the decline in Norwegian turnover in 2004 from 2003 is as much attributed to the rising value of the kroner against the US dollar (the currency in which the contracts are made) as it is to real reasons of business output.

Data sources

- Norwegian Space Centre (2002), *Annual Report 2001*, Oslo.
- Norwegian Space Centre (2006), *Annual Report 2005*, Oslo.
- Norwegian Space Centre (2006), *Space Economy, Industry, Indicators and Applications*, Oslo.

5.6. NORWAY

Figure 5.6a. **Turnover of Norwegian-produced space goods and services, 1997-2009**

Actual and projected (*) values in billions of Norwegian kroner

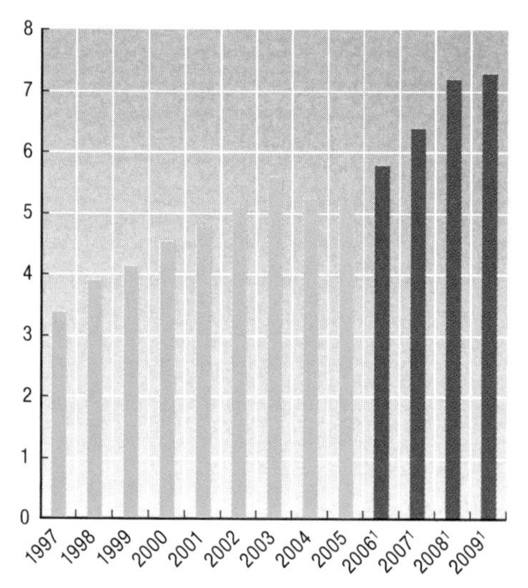

Source: Norwegian Space Centre (2006), *Annual Report 2005* , Oslo.

Figure 5.6b. **Export share as percentage of total Norwegian space-related turnover, 2002-2005**

Percentage of total turnover

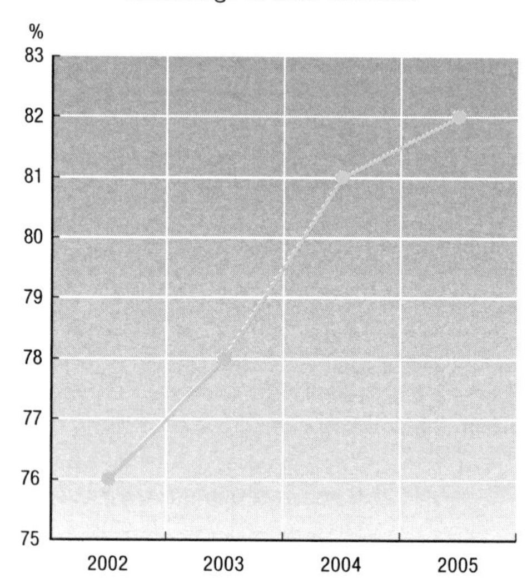

Source: Norwegian Space Centre, Annual Reports (2002-2005), Oslo

Figure 5.6c. **Spin-off effects factor for Norwegian ESA and NSC contracts, 1997-2009**

Actual and projected(*) spin-off multiplier: non-ESA sales per euro of ESA/NSC contracts

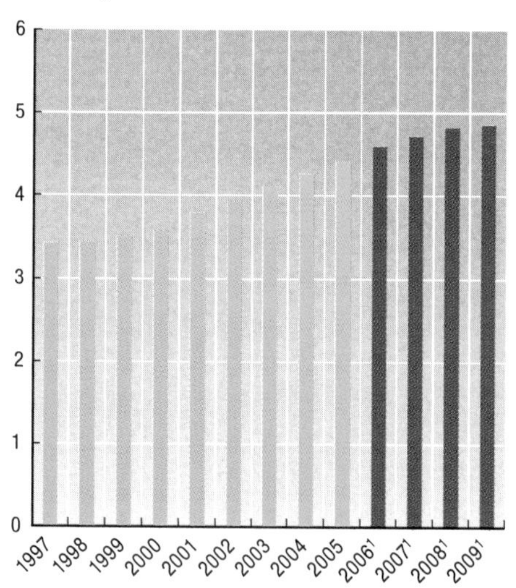

Source: Norwegian Space Centre (2006), *Annual Report 2005* , Oslo.

Figure 5.6d. **Total Norwegian ESA contracts and non-ESA spin-off sales, 2005**

Millions of Norwegian kroner

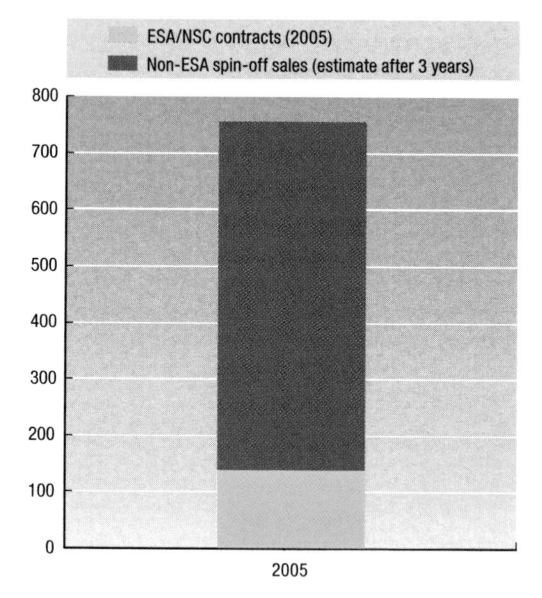

Source: Norwegian Space Centre (2006), *Annual Report 2005*, Oslo.

Annex A. The OECD Global Forum on Space Economics

In February 2006, the Organisation for Economic Co-operation and Development (OECD) launched a Global Forum on Space Economics under the aegis of the International Futures Programme (IFP). The purpose of the Forum is to help space agencies and governments to better identify statistically the sector and investigate its economic dimensions as an infrastructure for the larger economy.

Origins. Building on the experiences and recommendations of the OECD Futures Project "*The Commercialisation of Space and the Development of Space Infrastructure: The Role of Public and Private Actors*" (2002-2004), the Forum was born out of the need of a number of organisations for further economic analysis of the space sector, complementing the existing international platforms.

Objectives. The Forum will collect and evaluate data and socio-economic indicators. The aim is to provide evidence-based analysis for agencies and governments in shaping policies that contribute to realising the potential of space.

Participants. Participants in the Forum include the British National Space Center (BNSC), Centre National d'Etudes Spatiales (CNES), Canadian Space Agency (CSA), European Space Agency (ESA), Italian Space Agency (ASI), National Aeronautics and Space Administration (NASA), National Oceanic and Atmospheric Administration (NOAA), Norwegian Space Centre (NSC), US Geological Survey (USGS). Other agencies and ministries from OECD countries are expected to join. A companion Working Group is open to other interested parties and representatives of the private sector.

Activities. As of mid-2007, the Forum has undertaken three activities:

1. Dedicated work on statistics and economic indicators, to contribute to the emergence of internationally comparable data on the space sector and describe the sector's contribution to economic activity more broadly.
2. Work on "horizontal" case studies, which are meant to explore the broad economic and social dimensions of space applications (*e.g.* the first case study covers satellites' contributions to water management).
3. An annual update of the state of the sector with regards to the main OECD/IFP Space Project recommendations.*

* The Space Project's original recommendations are published in *Space 2030: Tackling Society's Challenges* (OECD, 2005).

Annex B. Case study: space technologies and water resources management

The management of the world's water resources is set to become one of the most important issues of the 21st century. In that context, and within the framework of the *OECD Forum on Space Economics*, the Forum Steering Group members commissioned an internal case study in 2006 to explore the general capabilities of space technology to enhance water resource management, and to see what space systems have generated so far in terms of socio-economic benefits and added-value in water management.[1] The findings briefly summarised here are mainly based on existing literature and meetings with practitioners, some of whose work is quite unrelated to the space sector.

The context

Significant strains on water worldwide. For decades now, the state of the world's water has given rise to concern. A number of international bodies, including the United Nations, the World Bank and the OECD, have warned policy makers for years of the daunting consequences of failing to adequately manage water resources. Population growth will continue to place huge strains on these resources. According to the *OECD Environmental Outlook* baseline, the global population will increase from slightly more than 6.1 billion in 2000 to over 8.2 billion in 2030. The demand for fresh water and sanitation will grow in parallel. Water withdrawals are expected to increase substantially by 2025: by 27% in developing countries and 11% in developed countries. This will put even more pressure on the search for new sources of water, on groundwater, etc. Economic growth and globalisation are expected to continue to grow apace, along with the population, thereby adding to the strains on water resources through increased demand and also increased pollution both of freshwater and coastal areas.

Extreme weather events. The social and economic impacts of increasing water demand and water pollution are difficult to estimate. Even more unpredictable are the effects of climate change. Extreme weather events, such as droughts, inland and coastal flooding, and hurricanes, are expected to increase significantly in coming years. The costs in terms of human lives and damage to economic assets and the environment could be huge. Hurricane Katrina for example cost insurers around USD 50 billion, but the total economic damage it inflicted could be as high as USD 200 billion. As globalisation progresses, large scale natural disasters in a given country could have huge economic repercussions in neighbouring countries, and also on commercial partners worldwide.

Role of space systems

Given such risk-laden long-term prospects, it is crucial to take action to adapt to and mitigate the effects. This will involve a range of different measures – some policy, some regulatory, some technical, etc. Within that range of instruments, space-based tools will have an important role to play.

An information infrastructure. A variety of satellites and ground systems already in place (though many still at the demonstration stage) are contributing significantly to several fields of application in water resources management. These systems range from meteorological satellites to Earth observation satellites for monitoring and measurement of specific Earth parameters, such as the bio-optical properties of oceans or water vapour. Space has become an increasingly important source of information, as ground-based monitoring systems have been lacking or have deteriorated in recent years. Data from meteorological satellites participate increasingly in operational water management, and key scientific discoveries have been made thanks to space-based data.

For example, the joint French-American mission Topex/Poseidon has used space altimetry to show that oceans have been rising over the past decade. Unexpected information has also been collected, such as variations in ocean circulations (*e.g.* El Niño, 1997-1998). These observations come at a cost that seems to be increasing as the number and length of missions increase, but a growing range of scientists and operational users request ever more data.[2]

Identifying benefits. But how are policy makers to decide on what level of financial, R&D and other resources to put into space to improve water resource management, and where to focus their efforts? The conventional approach to such questions is cost-benefit analysis. Numerous attempts have been made to measure the benefits of space-based systems more generally. But it has proven very difficult, if not impossible, to do so in a way that generates satisfactory results for the purpose of investment decision-making (See Table B1).

The difficulties also apply to attempts to quantify the use of space in water resource management. However, several socio-economic benefits have been identified, especially in terms of cost avoidance. Improved forecasts of El Niño in 1997-1998 – largely due to space systems data – are estimated to have saved California residents approximately USD 1 billion compared to the costs of a similar event in 1982-1983, which was not forecast.

Concerning a different water management application, the study *Real-time Ocean Services for Environment and Security* (ROSES) *Cost Benefit Analysis* developed a number of scenarios/options with certain assumptions to give an indication of the potential and prospective economic benefits a full oil spill detection system – using space assets – could provide in Europe by 2020 (Whitelaw *et al.* 2004.) This results in cost savings range from 1.5% to 2.25% of the oil cleanup bills of European countries, with potential benefits as high as EUR 12.4 million when the system is fully operational. As new operational systems to combat oil spills are progressively put in place using space-based imagery (*e.g.* in Canada, Norway, Italy), with the necessary surveillance mechanisms in parallel (*i.e.* deterrent aircraft patrols), useful new cost-efficiency studies will probably be conducted.

Investments: The risk-management approach

The rationale. Despite encouraging preliminary findings, the lack of accurately quantifiable benefits from the deployment of civilian space-based systems, coupled with the sheer unpredictability of many future events and their outcomes, clearly complicates major investment decisions. In light of this, it can be argued that policy makers need to explore new additional pathways to reaching decisions.

One such alternative is a risk management approach. The risks to human life and economic assets stemming from the effects of population growth, economic growth, globalisation and climate change on water resources are very substantial, and difficult to predict, and by the time they have happened, they may well be irreversible. In such circumstances, it makes eminently good sense to take action to better understand the risks, reduce uncertainty, reduce vulnerability to hazards, strengthen prevention, and improve the basis for mitigating the effects. In other words, the challenges facing the world's water resources need to be tackled through a kind of "insurance package" approach.

How much to invest. The question for policy making then becomes: what levels of premiums are appropriate? There would be understandable reluctance to pay excessive premiums. But are investments in a space infrastructure that help meet such objectives to be considered excessive? This would not seem to be the case. Comparing Earth observation and meteorological infrastructure with terrestrial infrastructures (roads, water, telecommunications), and bearing in mind the magnitude of potential losses in human life and economic assets, the overall cost of setting up such a system cannot be considered unduly high, nor are the rates of annual investment to maintain and expand the space infrastructure (see Section 2.2 on capital stocks).

Conclusion

With several OECD and non-OECD countries currently giving more importance to climate change policies, the need for adequate Earth observations will be more necessary then ever. Space infrastructure needs to be considered as a strategic asset in an infrastructure portfolio approach, where decision makers need to consider their options for improved risk management.

As shown in the case study, although the socio-economic benefits induced or derived from space-based infrastructure are difficult to assess, they do exist in many cases (*e.g.* key scientific advances, saved lives, economic activity derived from known water quality). More research needs to be conducted on this topic in the larger context of debates surrounding environmental valuation methodologies.

There will be increasing requirements for real-time monitoring and control at every point in the water cycle. Changes in current practice are required as managers in water resources need to make informed decisions about system changes in a "smarter" way. Advances in information technology and communications, coupled with space technologies such as Earth observation, could revolutionise their everyday use within the water sector, especially as the costs of the acquisition and use of these technologies drop.

Notes

1. The final version of this internal case study on water resources management (which will be complemented by another case study carried out from mid-2007 to early 2008) will be published in late 2008 by the OECD Global Forum on Space Economics.
2. Having access to nearly continuous data over long periods of time is very important for scientists to identify and analyse long-term climatic trends and changes.

Table B. **Main evaluation methods for the analysis of large programmes**

Method	Description	Comments for Space
Key performance indicators	Quantifiable performance measures.	Existing performance indicators at firm level. Move in some countries to develop new indicators in public agencies dealing with space.
Cost-benefit analysis	Measures tangible and intangible benefits and assesses these against costs.	Many studies conducted over the years for selected space applications with different methods, sometimes with inconclusive results.
Break-even analysis	The amount of time until benefits equal costs.	Not always applicable for large space programmes, where long-term investment in R&D may never be fully recovered. Method is used though in parallel to market studies for selected commercial applications, such as satellite telecommunications.
Transaction costs	Uses segmentation methods to calculate use and benefits to different user groups.	Not used much in the space sector so far, although the move to include more end-users in system development and funding (including operational users, such as data users in disaster management) will call for more transaction cost studies.
Cost effectiveness	Marginal costs for achieving specific goals.	Same as *Transaction costs.*
Value assessment	A complex method that captures and measures factors unaccounted for in traditional return on investment (ROI) calculations.	Challenging method, not used much so far, as it must include the difficult-to-measure costs and ROI of space and related ground systems (including long-term research and development).
Portfolio analysis	A complex method that quantifies aggregate risks relative to expected returns for a portfolio of initiatives.	Promising method that deserves to be more fully investigated according to OECD Forum on Space Economics.
Net present value	The difference between the present value of cash inflows and outflows at a given discount rate.	Method used, in parallel to market studies, in relatively mature commercial applications, such as satellite telecommunications.
Initial rate of return	The discount rate that makes net present value of all cash flows equal to zero	Same as *Net present value.*

Source: Adapted from OECD (2006), *Draft Report on Cost/Benefit Analysis of E-Government*, GOV/PGC/EGOV(2006)11, OECD, Paris.

Box B. **Tracking the world's water supplies**

Launched in March 2002, the Gravity Recovery and Climate Experiment (GRACE) is an international satellite mission to better understand the Earth's gravity field. Using a pair of satellites GRACE-1 and -2, which are 220 kilometres apart, water supply changes are measured around the world. Even if the water is captured in snow, rivers or underground aquifers, the satellites can detect the mass and trace its progress.

Scientific evidence indicates already that groundwater is being depleted in the central valley of California, parts of India, the central US, and in the Nubian Valley in Africa. So far, the data indicate that Africa in particular is losing a lot of water. The annual 21.6 millimetre losses between 2003 and 2006 in the Congo are, very roughly, equivalent to two years' worth of drinking water. Meanwhile, data from the GRACE mission show that the Nile has been going down an average of 9.3 millimetres a year while the Zambezi has declined by 16.3 millimetres. As natural climate variation can raise or lower water in a given period, observations need to be done over long periods to detect and monitor long-term problems.

Source: Based on Kanellos, M. (2006), "Satellites used to track world's water supply", CNET *News.com*, December 12.

Annex C.
General methodological notes

Purchasing power parity (PPP)

Comparing economies. In the early 1980s, the OECD and Eurostat established a programme to provide internationally comparable price and volume measures of GDP and its component expenditures for the member states of the European Union and the member countries of the OECD. This programme has since been enlarged with more than 40 countries contributing data. Before purchasing power parities (PPPs) became available, exchange rates had to be used to make international comparisons. But exchange rates do not reflect the relative purchasing powers of currencies in their national markets. Exchange rate converted data are generally misleading on the relative sizes of economies, overstating the size of economies with relatively high price levels and understating the size of economies with relatively low price levels. There is an additional problem that they are often subject to violent fluctuations. This means that countries may suddenly appear to become "richer" or "poorer", even though in reality there has been little or no change in the relative volumes of goods and services produced. Averaging exchange rates over several years dampens their fluctuations, but does not bring them closer to PPPs.[1]

How does it work? If the PPP for GDP between France and the United States is 0.97 euros to the dollar, it can be inferred that for every dollar spent on the GDP in the United States, 0.97 euros would have to be spent in France to purchase the same volume of goods and services. Purchasing the "same volume of goods and services" does not mean that identical baskets of goods and services will be purchased in both countries. The composition of the baskets will vary between countries and reflect differences in tastes, cultures, climates, income levels, price structures and product availability, but both baskets will, in principle, provide equivalent satisfaction or utility.

Limitations. PPPs are statistical constructs rather than precise measures. While they provide the best available estimate of the size of each country's economy and of the economic well-being of the country in relation to the others in the comparison, they are, like all statistics, point estimates lying within a range of estimates – the "error margin" – that includes the true value. PPPs are also an aggregate economic measure. Therefore, lower-level economic comparisons using PPPs are not always recommended, although they provide useful orders of magnitude. Cross-country productivity comparisons by industry, for example, should not be undertaken unless industry-specific PPPs are available.

Production

Production represents the value of goods and/or services produced in a year, whether sold or stocked. The related measure ***Turnover*** (not present in the OECD STAN database) corresponds to the actual sales in the year and can be greater than production in a given year if all production is sold together with stocks from previous years. While production and turnover will be different in a year, their averages over a long period of time should converge (depending on how perishable the stock is). Some care should be taken with the interpretation of production because it includes intermediate inputs (such as energy, materials and services required to produce final output, see Section 7 below on double counting). Any output of intermediate goods consumed within the same sector is also recorded as output, with the impact of such intra-sector flows depending on the coverage of the sector. For this reason, value added is often considered a better measure of output.

Business expenditure on R&D

Business enterprise R&D (BERD) covers R&D activities carried out in the business sector by performing firms and institutes regardless of the origin of funding. Industrial R&D is closely linked to the creation of new products and production techniques, as well as to a country's innovation efforts. The business enterprise sector includes firms, organisations and institutions whose primary activity is the market production of goods and services for sale to the general public at an economically significant price and the private and non-profit institutes that mainly serve them. The estimates presented in this book are from the OECD ANBERD database. This database was constructed to create a consistent data set that could overcome problems of international comparability and temporal discontinuities associated with the official BERD data provided to the OECD by its member countries. They are based on ISIC, Revision 3, from 1987 to 2000. This industrial classification only provides estimates for the aerospace industry as a whole.

Current and constant values

Current values. Current values (or values in nominal terms) represent the amount at that period of time when it was originally expensed or budgeted. The value is not adjusted for the effects of inflation or other price changes.

Constant values. Constant values (or values in real terms) indicate what an amount would be worth in comparison to some value from a base period. This is done mainly to eliminate price differences from year to year and ease comparisons. The most common way uses an index, particularly the Consumer Price Index (CPI), to convert the values into that of the base year. CPI is a measure that examines how the price of a weighted-basket of goods and services purchased in the basic economy varies over time.

Limitations with constant values and space. A particular problem in trying to put the space sector's values into constant dollars is trying to determine which index to use when trying to put values into a base year for space products or services. CPI measures change in the value of goods and services in the general economy, but so far no price indices presently exist that specifically relate to space items. In that context, prices from year to year may vary considerably due to a number of factors: different specifications, varying countries, dual use *versus* single usage and other issues.

Example. The government of a country has R&D expenditures of USD 100 in 2005 and 2006. The current value would be USD 100 in both years, regardless of the inflation rate, as that was the actual amount of expenditure for those respective years. With constant prices, the value spent in 2005 (the base year) would be USD 100. In the next year, 2006, and assuming an inflation rate of 10%, the original USD 100 expenditures would be worth only USD 90 (100/

(1+r)), where r is the inflation rate. The USD 10 reduction in 2006 constant values indicate that USD 100 in 2006 would only be able to buy USD 90 worth of items from 2005.

Nominal and real exchange rates

Exchange rates indicate the value of one currency in terms of another. One of the most common methods is to provide the exchange rates without adjusting them for the price differences that may exist between the countries involved; these are known as nominal exchange rates.

Conversely, real exchange rates try to put the values of the currencies in terms of their price levels to try to reflect more appropriately what a good or service in one country may trade for in another. One of the most popular ways to try to estimate the real exchange rate is to use purchasing power parities (see information on PPP above).

Productivity

Productivity is a measure of efficiency and represents the amount of output that is derived for a given amount of input. Productivity can be measured in physical (*e.g.* total number of units produced per unit of input) or financial (*e.g.* the US dollar value of outputs per unit of input) terms. One of the most commonly used types is labour productivity, which measures the value of output provided for a given unit of labour (*e.g.* usually measured in terms of either per worker or per labour-hour). However in very highly technical industries, such as the space manufacturing industry, much of the increase in labour productivity may be as attributable to newer technologies (*i.e.* machines) or better management practices, as to a better trained and educated workforce.

Double counting

Double counting is a frequent problem when examining statistics. It is an accounting error in which an item is taken into account more than once. This is a particular concern when looking at production data. Production represents the value of goods and services produced by an entity. However, if one enterprise produces an item for USD 100, which is than sold to another enterprise in the same industry that produces USD 500 of goods, there is a need to identify the USD 100 item separately when analysing the global value created or run the risk of "double counting". The USD 100 component will appear in the production value of both the originating manufacturer and the company including that component into its product. One way to avoid this is to use value added, which may not always be as easily readable or accurate, but will exclude the USD 100 component in measuring the output of the enterprise purchasing it from the original manufacturer. Similar, problems may also exist when examining other variables such as sales.

Notes

1. See OECD (2005), *Purchasing Power Parities and Real Expenditures: 2002 Benchmark Year.*

Annex D. Space-related statistics from OECD sources

Various OECD internal sources* have been used to draft this report.

- **The OECD STructural ANalysis (STAN) database** provides a comprehensive tool for analysing industrial performance at a relatively detailed level of activity across countries. It includes annual measures of output, labour input, investment and international trade which allow users to construct a wide range of indicators to focus on areas such as productivity growth, competitiveness and general structural change. Through the use of a standard industry list, comparisons can be made across countries. STAN is primarily based on member countries' annual national accounts by activity tables and uses data from other sources, such as national industrial surveys/censuses, to estimate any missing detail. Since many of the data points in STAN are estimated, they do not represent official member country submissions. STAN is based on the International Standard Industrial Classification of all economic activities, Revision 3 (ISIC Rev. 3) and covers all activities (including services). The latest version of the database was released at the end of 2005 (with data up to 2003) and work on a new and more extensive STAN data system is underway.
- **The STAN Bilateral Trade Database (BTD)** is compiled by the Economic Analysis and Statistics Division (EAS) of the Directorate for Science, Technology and Industry (STI). This database is designed to provide analysts with information on exports and imports of goods in OECD countries, broken down by partner country (or geographical area) and by economic activity. BTD is derived from the OECD's International Trade by Commodity Statistics (ITCS) database, where (values and quantities of) imports and exports are compiled according to product classifications and presented by partner country.
- **The ANBERD (Analytical Business Enterprise Research and Development) database** was developed to provide a consistent data set that overcomes the problems of international comparability and breaks in the time series of the official business enterprise R&D provided to the OECD by its member countries through the OECD's R&D survey. Through the use of established estimation techniques, the OECD Secretariat has created a database for 19 of the largest R&D performing countries, as well as a zone total for the European Union. The database is designed to provide analysts with

* More information on the different OECD databases can be found on the OECD Website at *www.oecd.org*.

comprehensive and internationally comparable time-series on industrial R&D expenditures.

- **OECD and Patents Statistics.** OECD work on patent statistics is conducted in close co-operation with the members of the *Patents Task Force*, which brings together the world's major patent offices (European Patent Office, Japanese Patent Office, United States Patent and Trademark Office and the World Intellectual Property Organisation), as well as major providers of statistics and indicators on science and technology (the European Commission, Eurostat, and US National Science Foundation). The main objective is to develop an international statistical infrastructure for patents, with a strong emphasis on the development of databases and methodologies. This infrastructure provides the conditions for improving the comparability and quality of patent indicators, enhancing the accessibility of patent statistics and facilitating the development of a new generation of indicators for policy and research use.

OECD PUBLICATIONS, 2, rue André-Pascal, 75775 PARIS CEDEX 16
PRINTED IN FRANCE
(03 2007 02 1P) ISBN 978-92-64-03109-8 – No. 55397 2007